# MAKING WHEELS

# MAKING WHEELS

*A technical manual on wheel manufacture*

R.A. DENNIS

INTERMEDIATE TECHNOLOGY PUBLICATIONS 1994

Published by ITDG Publishing
103–105 Southampton Row, London WC1B 4HL, UK
www.itdgpublishing.org.uk



First published in 1994
Print on demand since 2003

ISBN 1 85339 141 7



A catalogue record for this book is available from the British Library

ITDG Publishing is the publishing arm of the Intermediate Technology Development Group.
Our mission is to build the skills and capacity of people in developing countries through the dissemination of information in all forms, enabling them to improve the quality of their lives and that of future generations.

Printed in Great Britain by Lightning Source, Milton Keynes

# CONTENTS

1. **Introduction** . . . . . . . . . . . . . . . . . . . . . . . . . . . . . . . . . 1

1.1 Background, 1
1.2 The wheel manufacturing technology, 2
1.3 Content of the manual, 3
1.4 Use of the manual, 4
1.5 Summary of specifications, 5

2. **Rim-bending machine** . . . . . . . . . . . . . . . . . . . . . . . . . 12

2.1 Introduction, 12
2.2 Construction of rim-bending devices, 15
2.3 Operation of the rim-bending machine, 36

3. **Assembly jig** . . . . . . . . . . . . . . . . . . . . . . . . . . . . . . 43

3.1 Introduction, 43
3.2 Construction of assembly jig, 45
3.3 Use of the assembly jig, 61

4. **Wheel-axle assemblies** . . . . . . . . . . . . . . . . . . . . . . . . 65

4.1 Bearing/axle assemblies, 65
4.2 Types of bearings, 69
4.3 Selection of bearing type, 71
4.4 Axle design, 75
4.5 Axial thrust washers, 76
4.6 Sealing of bearings, 78

5. **Manufacture of wheels to take bicycle and motorcycle tyres** . . . . . . . . . . . . . . . . . . . . . . . . . . . 83

5.1 Design of wheels to take bicycle tyres, 84
5.2 Construction of hub axle assembly, 87
5.3 Construction of bicycle-type wheels, 89
5.4 Manufacture of a wheel to take a motorcycle tyre, 96

6. **Wheels for motor vehicle tyres** . . . . . . . . . . . . . . . . . . . 100

6.1 Wheel designs, 101
6.2 Hub/axle assembly, 106
6.3 Construction of wheels, 123
6.4 Construction of detachable-bead wheel to take a motorcycle tyre, 135

**Appendix 1** Construction of a hand-operated machine for boring out hardwood bearing blocks, 139

**Appendix 2** Technical drawings of the rim-bending machine, 145

# ACKNOWLEDGEMENTS

This manual and the technology presented in the manual have been developed within the Rural Transport Programme of the Intermediate Technology Development Group.

ITDG gratefully acknowledge donations from the following companies - Burmah Castrol Trading Ltd., MCL Group Ltd., NFC Plc., and Powell Duffryn Plc., - which have helped to fund the production of the manual.

The author also gratefully acknowledges the contributions of other technical staff in I.T. Transport Ltd., Alan Brewis, John Lees and Alan Smith, to the development of the technology and of Mrs. Lynn Curtis to the preparation of the manual. Thanks are also due to Jeremy Hartley and Alan Spence for additional photographs.

# 1. INTRODUCTION

## 1.1 BACKGROUND

Up to 80 or 90% of goods transport in rural areas of developing countries is still carried out on foot. This imposes a considerable burden in both time and effort on rural communities and is a considerable constraint on improving standards of living. Simple wheeled vehicles such as handcarts, bicycle-trailers and animal-drawn carts can have a major impact on reducing the transport burden but at present they are used for only about 10 to 15% of goods transport. Although limited affordability and lack of credit facilities are usually the main limitations on their wider use, problems with the production and supply of these vehicles are also often serious constraints.

The main problem in producing low-cost vehicles, particularly in rural areas, is usually the lack of suitable wheels and axles. The quality of the wheel-axle assembly is the most important factor affecting the performance of simple vehicles and the wheel-axle is also the main cost item, usually comprising at least 50% of total cost. Assisting workshops to overcome problems with the supply of wheel-axle assemblies and to upgrade the quality of these items therefore has the potential to provide major benefits for rural communities in increasing the availability of appropriate vehicles, possibly at lower cost.

The high capital-cost, large production methods used to produce wheels in industrialised countries are generally not appropriate for developing countries, nor is it feasible to use imported wheels. Scrap wheel-axle assemblies from motorised vehicles are quite widely used but in general the supply of these is not adequate or reliable enough to sustain the required level of production of low-cost vehicles. Some large workshops may produce wheels for animal-drawn carts and low-speed trailers on high capacity presses but most locally produced wheels are either made from wood or fabricated from steel sections.

Large wooden wheels are widely used in Asian countries where the skills needed to construct these types of wheels have been developed over many years. However, the high quality woods needed to make good quality wheels are becoming increasingly scarce and these wheels are becoming more expensive. There is also increasing concern about the damage done to roads by these wheels. There have generally not been the same traditions for wooden wheel building in African countries and consequently wooden wheels here tend to be fairly crude and not to perform well so that they are not widely acceptable.Other problems with wooden wheels are that they are heavy, prone to damage from both operation and the weather, and they can only be used with steel or solid rubber tyres.

The common type of fabricated steel wheel produced in developing countries has a rim bent or rolled from flat bar and spokes cut from flat, round or angle section. The wheels are heavy and often suffer from failure of the welds between the rim and spokes. The rim may run directly on the ground or may be fitted with a strip of rubber - often cut from a scrap tyre. Bending of the rims is often a major problem because of the limited equipment available and consequently the wheels are often fairly crudely constructed and their performance is generally poor. The quality and performance of these wheels could be greatly improved if more structurally efficient sections were used for the rim, better equipment was available to form the rims, and rubber tyres were used which give some degree of cushioning on rough roads.

The wheel manufacturing technology described in this manual has been developed in response to the situation outlined above. Experience with low-cost means of transport in over 30 developing countries over a period of more than 12 years has clearly shown the need to provide small to medium sized workshops with a simple technology for producing good quality wheel-axle assemblies in order to improve their output of efficient, low-cost vehicles. It is considered that localised production in smaller workshops is more appropriate than large-scale, centralised production as these workshops are more closely integrated with their local communities and are better able to respond to local needs for both supply and maintenance of vehicles. However, experience has indicated that some degree of specialised manufacture may be desirable in an area to achieve the best standards of production with certain more capable workshops being encouraged to concentrate on producing wheels and axles to supply to other workshops.

## 1.2 THE WHEEL MANUFACTURING TECHNOLOGY

The technology comprises a hand-operated bending machine which can bend a range of steel sections into good quality wheel rims of various sizes and an assembly jig to set up the wheels for welding. The rim-bender works on the principle of feeding the rim material through the bender in small lengths and bending it continuously so that an accurate circle is produced. A large leverage is used so that quite stiff sections can be bent without too much effort. The bender can be continuously adjusted so that any rim diameter greater than a minimum of 300mm can be formed. The rim and hub are clamped in the assembly jig so that the wheel is accurately made with the hub exactly at the centre of the rim. The assembly jig therefore ensures that accurate wheels of consistent good quality are obtained.

The wheel manufacturing equipment has been designed to allow construction in workshops with basic metal-working skills and tools. The design minimises the number of steel sections that are needed and attempts to use only sections that are commonly available. It is also flexible so that, wherever possible, alternative sections can be substituted for materials that are not available. The basic design presented in the manual can be constructed with tools for cutting, drilling and welding steel, an angle-grinder is also desirable for grinding some of the welds flat. Hand tools can be used for cutting but obviously a power-saw will save considerable effort. The maximum hole size needed is 12mm which can be drilled with a hand-drill but again a pedestal drill saves effort. Most of the sections used are 6mm thick flat bar and angle, and arc-welding is relatively straight forward. However, since accuracy, and for the bender, strength, are very important, good welding skills are needed.

The major skill needed in making the equipment is accuracy in cutting and setting up the pieces. Squareness and alignment of parts is very important for the equipment to work properly, and in the bender, parts must fit together closely but still be able to slide and rotate freely. A methodical and careful approach is needed to check and recheck the accuracy and alignment of parts.

The manufacture of wheels with the equipment requires the same basic metal-working skills and tools as above but the accuracy of construction is less demanding. Experience has shown that the equipment is best made in a workshop with a good level of capability but that most small workshops are able to use the equipment to make good quality wheels.

The wheel manufacturing technology has been developed and tested over a period of about eight years. It has now been introduced into over sixty workshops in ten countries and has clearly been proven to be appropriate for use in small to medium size workshops. It enables them to have their own facility for producing a range of low-cost wheels to take various types of tyres and to suit all types of low-speed vehicles.

### 1.3 CONTENT OF THE MANUAL

The manual contains a set of step-by-step instructions for construction of the wheel rim bending machine and of the assembly jig. These are reinforced by a set of engineering drawings of the bender contained in an appendix at the end of the manual. It also contains step-by-step instructions for operation of the equipment and for the construction of a number of basic wheel designs using the equipment.

The instructions are well illustrated with both two-dimensional and pictorial sketches so that it is not necessary to be able to read technical drawings to make use of the manual. The sketches are supported by photographs to show what the finished components should look like and photographs are also used to summarise the steps in making the wheels.

Although the instructions specify preferred sizes of materials, efforts have been made to keep the designs flexible so that wherever possible alternative materials may be used. Hopefully, therefore, construction of the equipment and/or wheels should not be prevented by the unavailability of certain specified sizes of material.

As stated above, the rim bending machine can be made with basic metalworking equipment. However, a slightly neater and simpler version can be made if a drilling machine is available that can drill holes of at least 32mm diameter and details are also given of the construction of this alternative version.

Two basic types of wheel design are described in the manual:

1. A bicycle-type wheel which can be fitted with standard bicycle tyres or solid rubber tyres. This wheel has an angle section rim and spokes which may either be angle section or solid round bar. The wheel is mainly for use on cycle trailers and light handcarts. A similar design using heavier sections for the rim and spokes and fitted with motorcycle or solid rubber tyres can be made for use on heavy-duty handcarts.

2. A wheel which may be fitted with car or motorcycle tyres. The wheel may have a split-rim or a detachable bead so that the tyre may be easily slid onto the main part of the rim and then held in position by bolting on the other part of the rim or the detachable bead. The tyre can therefore be readily fitted with little effort and with only the use of a spanner. This type of wheel is intended for use on animal-drawn carts and low speed trailers.

Some of the wheels produced with the equipment are shown in Figure 1.1.

Both types of wheels have been extensively tested on a wheel test rig to make sure that they are strong enough for their intended use on rough earth roads.

Details are also given of a number of bearing and axle assemblies which can be used with the wheels so that the manual covers the manufacture of the complete wheel and axle assembly.

The layout of the manual is as follows:

**Chapter 2** contains the instructions for construction and operation of the rim-bending machine.

**Chapter 3** contains the instructions for construction and use of the assembly jig.

**Chapter 4** gives a general discussion on the design of axle assemblies and the selection of various types of bearings.

**Chapter 5** describes the manufacture of wheels with angle-section rims to take bicycle and motorcycle tyres and also solid rubber tyres.

**Chapter 6** describes the manufacture of split-rim and detachable bead wheels to take car and motorcycle tyres.

**Appendices** contain details of the construction of a hand-operated machine for boring out hardwood bearings and also the set of working drawings of the rim-bending machine.

## 1.4 USE OF THE MANUAL

A draft version of the manual has been widely field tested over the past two years and this final version contains changes and additions suggested by feedback from the use of the draft version.

Experience from the field trials has shown that workshops with basic metalworking skills and equipment can construct the wheel manufacturing equipment from the manual but that a reasonable level of competence is needed to construct it to the standard needed for it to work effectively. Workshops with qualified staff can construct the rim bending machine from the engineering drawings included in Appendix 2, but it is highly recommended that the step-by-step instructions given in Chapter 2 are consulted, as these point out important aims of the construction process and ways in which these can be achieved. All drawings for the construction of the assembly jig are included in the step-by-step instructions given in Chapter 3 and it is highly recommended that these instructions are closely followed in order to achieve the level of accuracy needed for the jig to be effective.

Chapter 4 presents a general discussion of hub and axle assemblies which are suitable for the wheel designs described in the manual. It is intended for readers who are interested in the general aspects of wheel design. It also provides additional explanation of the specific hub and axle designs presented in Chapters 5 and 6.

Details of specific wheel designs are given in Chapters 5 and 6 and these are the chapters which will be most useful to workshops that wish to use the equipment to produce wheels. These chapters contain both details of designs and also step-by-step instructions for construction of the wheels. More general explanations of the use of the rim bender and assembly jig for manufacturing wheels are included in Sections 2.3 and 3.3 respectively.

Although the use of the manual is the most cost-effective method of introducing and disseminating the technology, the reading, interpretation and implementation of the manual obviously involves a fair degree of effort and there is a chance that there will be some points of uncertainty. Additional inputs which can help to make the dissemination more effective are as follows:

1. A wheel manufacturing kit and sample wheels can be supplied from Zimbabwe. This will cost about US$300 plus packing and freight costs. Many workshops find it very helpful to have samples of the hardware to see what they are making and to show the quality of construction needed.

2. A training course can be organised. If this is in-country about six to eight workshops would need to participate to make it worthwhile. The course, lasting two weeks, would cover the use of the equipment to make a number of wheels and also the construction of one or two low-cost vehicles incorporating the wheels, such as cycle trailers or wheelbarrows. An "in-country" course would be likely to cost between US$6000 and $9000, which would include supply of locally made kits to the workshops. Alternatively, it may be possible to arrange a training course for one or two people in a regional training centre. The cost of this, including intra-country airfares, would be about US$1500 to $2000 per person.

In both these cases the manual would supplement the additional inputs and provide the workshops with a record of all the information needed to make and use the equipment.

## 1.5 SUMMARY OF SPECIFICATIONS

This section lists the main details of the construction of the wheel-making equipment and of the wheel designs included in the manual.

## Wheel rim bender

### i) Basic Model:

**Materials:**

| | | |
|---|---|---|
| Angle - | 40 x 40 x 6 | 3.8m |
| Flat bar - | 75 (or 80) x 6<br>25 x 6 | 1.5m<br>1.0m |
| Round bar -<br>for pins | 38 or 40 diameter<br>Bright Mild Steel (shafting) | <br>0.95m |
| Threaded rod - | M20 | 0.15m |
| Steel pipe - | 2" medium wall | 0.15m |

**Skills and equipment:**

welding; drilling up to 10mm diameter; hand grinder; hand tools for measuring, cutting and accurately setting up parts.

**Time for construction:**

4 to 5 days

### ii) Alternative Model

**Materials:**

| | | |
|---|---|---|
| Angle - | 40 x 40 x 6 | 1.6m |
| Flat bar - | 65 x 12<br>25 x 6 | 2.0m<br>1.0m |
| Round bar -<br>for pins | 32 diameter<br>38 or 40 diameter<br>Bright Mild Steel | 0.3m<br><br>0.6m |
| Threaded rod - | M20 | 0.15m |
| Steel pipe - | 2" medium wall | 0.15m |

**Skills and Equipment:**

welding; drilling up to 32mm diameter; hand grinder; hand tools for measuring, cutting and accurately setting up parts.

**Time for construction:**

3 to 4 days

**Assembly Jig:**

**Materials:**

| | | |
|---|---|---|
| Angle - | 50 x 50 x 6 | 4.5m |
| Flat bar - | 50 x 6<br>100 x 6 | 2.5m<br>0.1m |
| Round bar - | 16 or 20 diameter | 0.7m |

M16 threaded rod (1.0m) and 5/8" medium wall pipe (0.5m)

**or** M20 threaded rod (1.0m) and 3/4" medium wall pipe (0.5m)

These are the main materials which are used in the designs but in most cases details are given of alternative sizes which can be used.

**Skills and equipment:**

Welding; drilling up to 12mm diameter (16 or 20mm diameter would be better); hand grinder; hand tools for measuring, cutting and accurately setting up parts access to lathe needed to machine locating collars for wheel hubs.

These are the basic needs - obviously time and effort will be saved if other machine tools are available, for instance a power hacksaw.

**Time for construction:**

4 to 5 days

**Cost:**

the above information should allow a good estimate to be made of the cost of producing the wheel manufacturing equipment. As a guide, in Zimbabwe the cost of having the equipment (bender and assembly jig) made up in a commercial workshop is about US$280.

## Details of wheels included in manual

**1. Wheels to take bicycle tyres:**

i) Angle-section rim and angle section spokes - Figure 5.1

Material: 25 x 25 x 3 angle - 4.2m

ii) Angle-section rim and round bar spokes - Figure 5.2

| Materials: | 25 x 25 x 3 angle - | 2.4m |
|---|---|---|
| | 8 diameter reinforcing rod - | 3.5m |

**Hubs and axles:**

3/8" bicycle axle with bearing cups and balls; 1" medium-wall pipe hub (8cm long)

**or** 25mm diameter stub axle with 6205 ball bearings; 2" medium-wall pipe hub (8cm)

**or** 1" medium-wall pipe live axle with hardwood block bearings

**2. Wheels to take motorcycle tyres:**

i) Angle section rim - Figure 5.5

| Materials: | Angle 40 x 40 x 3 - | 1.7m |
|---|---|---|
| | Angle 25 x 25 x 3 - | 1.2m |

ii) Detachable bead wheel - Figure 6.11

| Materials: | Flat bar 50 to 60 wide x 3 to 5 thick - | 1.35m |
|---|---|---|
| | 10 or 12 diameter round bar - | 3.2m |
| | Angle 25 x 25 x 3 - | 1.8m |

**Hubs and Axles:**

25mm diameter stub axle with 6205 ball bearings; 2" medium-wall pipe hub (8cm)

**or** 1 1/2" medium-wall pipe live axle with hardwood block bearings

## 3. Wheels to take car and truck tyres:

i) Split-rim wheel - Figures 6.1 and 6.2

Materials for 16" tyre:

| | |
|---|---|
| Flat bar 65 to 80 wide x 5 or 6 thick - | 1.25m |
| 40 to 65 wide x 5 or 6 thick - | 1.25m |
| 16 diameter round bar - | 3.1m |
| Angle section 50 x 50 x 6 - | 0.8m |

ii) Detachable bead wheel - Figure 6.3

Materials for 16" tyre:

| | | |
|---|---|---|
| Flat bar - | 100 x 5 or 6 thick - | 1.25m |
| | 50 x 5 or 6 thick - | 0.3m |
| 16 diameter round bar - | | 3.1m |
| Angle section 50 x 50 x 6 - | | 0.9m |

**Hubs and Axles:**

40mm diameter stub axle with 6208 ball bearings or taper roller bearings;
3" medium-wall pipe hub (10.5cm)

**or** 2" medium-wall pipe live axle with hardwood block bearings

**Notes:**

### 1. Skills and equipment:

Basic metal working skills and equipment for welding, drilling hole sizes up to 10mm, hand grinding, and hand tools for measuring, cutting and setting up parts. A lathe would be useful for machining hubs and axles but is not essential.

### 2. Time for construction:

On a "one-off" production basis, a welder and assistant could produce a wheel in about 3/4 of a day. However, it is better to produce the wheels on a batch process of say 10 or 20 wheels at a time. In this way it has been shown by experience that about 10 wheels can be produced per week.

### 3. Cost:

The cost of materials and labour can be estimated from the information given above. Other main costs will be of tyres, inner tubes and bearings.

### 4. Alternative designs:

Although details are given of designs for specific sizes of pneumatic tyres, data is also given for construction of similar wheels for other tyre sizes. The designs can also be readily adapted to take solid rubber tyres.

Ox-cart with split-rim wheels

Ox-drawn water cart with detachable bead wheels

Wheelbarrow wheel with angle-section rim and solid rubber tyre

**Figure 1.1: EXAMPLES OF WHEELS PRODUCED WITH THE WHEEL MANUFACTURING TECHNOLOGY**

Handcart with split-rim wheels fitted with scrap motorcycle tyres

Cycle trailer with wheels fitted with 28" bicycle tyres

Cycle-ambulance with wheels fitted with 28" bicycle tyres

# 2. RIM-BENDING MACHINE

## 2.1 INTRODUCTION

This is a simple, hand-operated device which can bend a variety of steel sections into good quality wheel rims (or circular shapes required for other purposes). Details of the device are illustrated in Figure 2.1 and in the drawings in Appendix 2.

It comprises two lower rollers and a central, upper forming tool which is mounted in a tool holder attached to a lever arm. Three different forming tools may be used to cover the range of steel sections which will normally be bent in the machine. Each forming tool is fixed to a tool pin which slides into the bottom end of the tool holder.

The tool pin slides in vertical guides inside the main frame. The upper end of the tool holder is connected by a pin to the lever arm which in turn is pivoted in the main frame. The lever arm has a socket for an extension arm which is designed to take 1 1/2" pipe. Various lengths of extension arms may be used depending on the sections to be bent and the operating forces needed.

There are two positions for the lower rollers:

- the OUTER position should be used **whenever possible** as the forming forces are lower for this position, thus causing less wear and tear on the machine and minimising the risk of distorting the sections being bent;

- the INNER position is used for bending smaller diameter circles. It is possible that some flattening of sections such as angle and tube may occur with the rollers in this position. However, this problem is unlikely to arise in the manufacture of wheel rims.

The degree of bending and diameter of circle obtained is controlled by the adjustable stop on the side of the device. This controls the movement of the lever arm and as it is screwed downwards allows more bending to be applied. A guide to the adjustment of the stop is given in Section 2.3. It should be noted that the actual curvature obtained depends on the spring back of the bar after bending and therefore on the material being bent. Some degree of trial and error will therefore normally be needed to get an exact curvature.

The bending machine described in the manual has an inside frame width of 106mm i.e. the maximum width of bar that can be bent is 105mm. If a machine is needed to bend larger widths it is quite straightforward to increase the width of the tool pin and main frame accordingly.

**Alternative Design of Bending Machine**

The drawings in Appendix 2 and the step-by-step instructions in Section 2.2 describe the construction of a rim bender which can be made in a workshop with basic facilities for cutting, welding, grinding and drilling hole sizes up to 12mm. Therefore, the large diameter pivot pins are fitted into housings fabricated from angle section.

If the workshop has a drilling machine for drilling hole sizes of at least 32mm a simpler and neater design can be constructed in which the pivot pins are fitted into machined holes. This design is shown at the end of Section 2.2. The construction follows the same steps as outlined for the basic design.

**Capacity of the Bending Device**

The following figures give an approximate indication of the capacity of the device:

| Type of section | Maximum size that can readily be formed | Minimum diameter that can be formed |
|---|---|---|
| 1. Rollers in OUTER position (266mm centres) | | |
| • flat | 100 x 12mm | 450-500mm |
| • angle | 40 x 40 x 6mm | 400-450mm |
| • round | 25mm diameter | Approx. 400mm |
| • tube | 3/4" water pipe | Approx. 400mm |
| 2. Rollers in INNER position (154mm centres) | | |
| • flat | 100 x 6mm | Approx. 300mm |
| • angle | (tends to flatten - capacity needs to be found by experiment) | |
| • round | 16mm diameter | Approx. 300mm |
| • tube | 5/8" water pipe | Approx. 300mm |

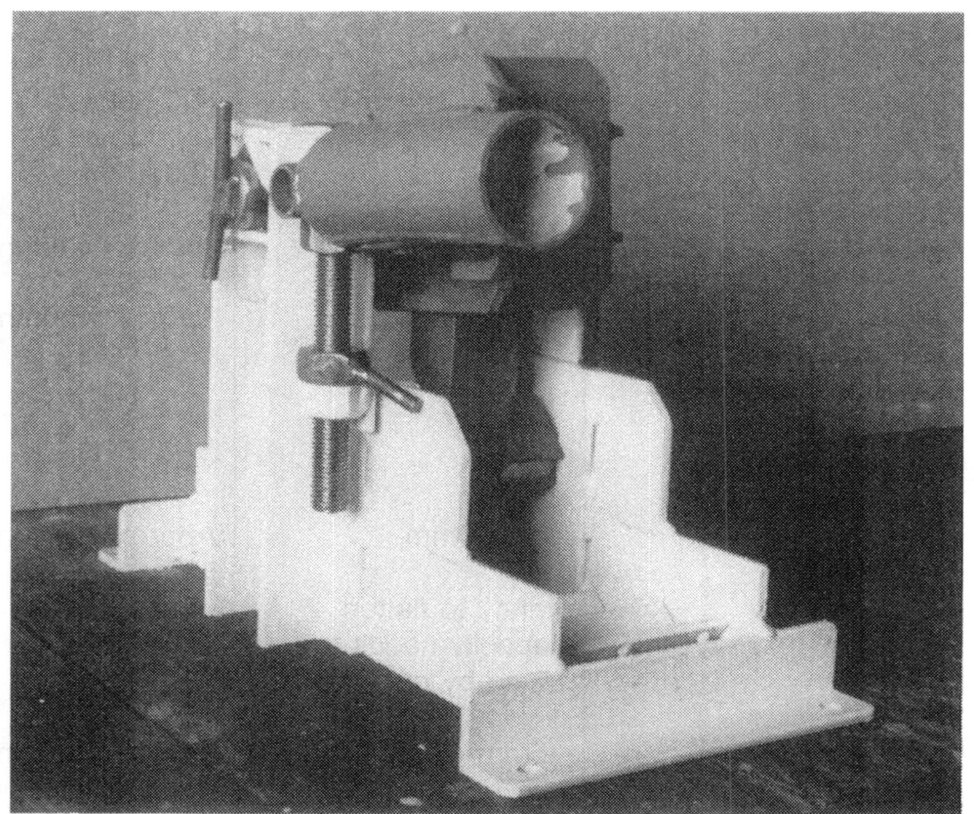

**Figure 2.1: RIM-BENDING MACHINE**

## 2.2 CONSTRUCTION OF RIM-BENDING DEVICES

The following set of sketches and notes give a step-by-step sequence of instructions for construction of the rim-bending device. The steps have been planned to ensure, as far as possible, a straightforward and accurate assembly of the device.

**It is emphasised that accurate measurement, cutting and setting up of the parts is very important to achieve the best performance of the bending device. A good quality of welding is also needed to minimise distortion of the parts and to ensure that the device is adequately robust and durable.**

Measurement and cutting should be to an accuracy of + 1mm and a good quality **square** should be used to ensure that parts are properly set up square. **It is particularly important that the two sides of the frame are vertical and correctly aligned so that pins fit correctly and that the axes of all pins are parallel** Pins should be a **good fit** in their housings to minimise `play´ in the mechanism - excessive clearances may reduce the smoothness of operation of the device and lead to some inaccuracy in the consistency of bending.

**Note:** the step-by-step instructions should be used in conjunction with the set of working drawings provided at the end of this manual.

### Materials

The design attempts to minimise the number of different sizes of steel section which are used. The basic sections are as follows:

| Section | Preferred size | Length needed |
|---|---|---|
| Angle | 40 x 40 x 6 | 380cm |
| Flat Bar | 75 x 6<br>25 x 6 | 150cm<br>100cm |
| Round Bar for Pins | 38 or 40 diameter<br>Bright bar (shafting) | 95cm |
| Threaded Rod | M20 | 15cm |
| Steel Pipe | 2" medium-wall | 15cm |

**Angle section:** 40 x 40 x 6mm is the basic section used throughout and it is best to keep to this section if at all possible. To maintain strength and rigidity a smaller section should not be used.

If a larger section (say 50 x 50 x 6mm) is used the frame will need to be wider to accommodate a wider tool holder and lever arm. The short lengths making up the housings for the pins in the frame and lever arm should be trimmed down to size to suit the fitting of the pins.

**Flat Bar:** 2 sizes are used, 75 x 6 and 25 x 6 - alternative widths are possible but the 6mm thickness should be maintained if possible. A smaller thickness should **not** be used - if a greater thickness is used (e.g. 8mm) the width of the frame and lever arm must be increased to maintain the basic throat width of 106mm.

**Round Bar:** The **preferred** size for all pins is 38 or 40mm diameter. **Cold drawn** (bright) steel bar (sometimes known as shafting) should be used if available. 32, 35 or 45mm diameter bar may be used as an alternative but the lever arm will need to be modified to suit.

## 1. FRAME

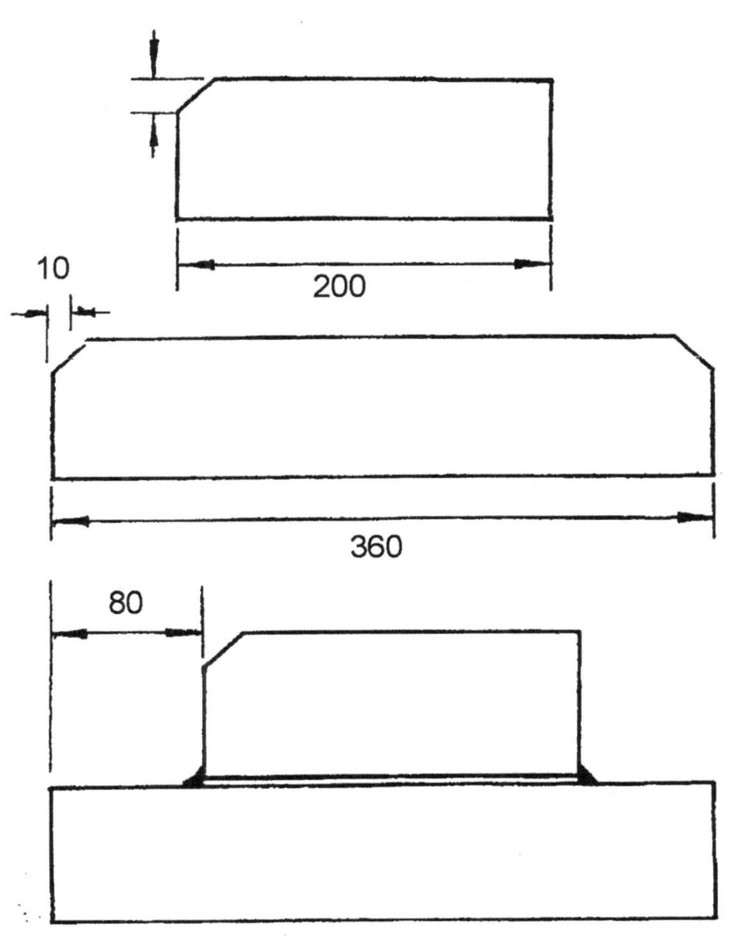


**Step 1: Construct sides of frame**

**1.1:** Cut **2** pieces of each size shown from **6** or **8**mm thick **bar**. Preferred width is **75**mm, but other widths from **50** to **80**mm may be used.

Trim corners as shown.

**1.2:** Line up each pair of pieces on a flat surface. Leave a gap of about 2mm for welding. Hold pieces in position and tack-weld at corners.

**Step 2: Fit 2 angle uprights to each side**

**2.1:** Cut **4** lengths of **40 x 40 x 6 angle** of **exactly** the same length.

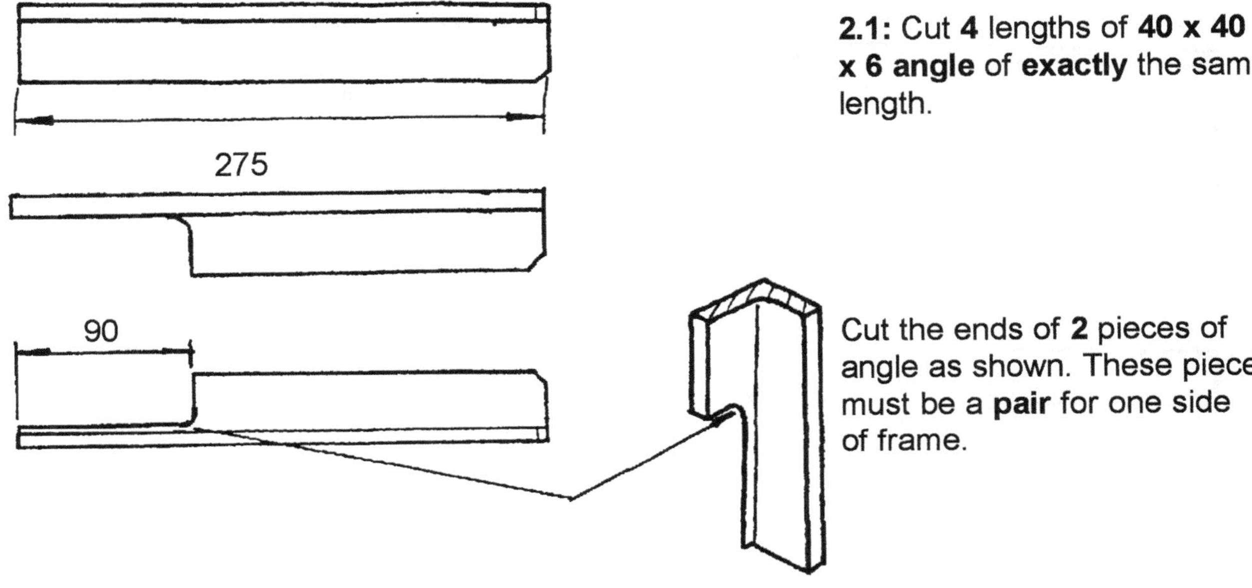


Cut the ends of **2** pieces of angle as shown. These pieces must be a **pair** for one side of frame.

**Step 3: Attach tool-pin guide strips on inside of side plates**

**3.1:** Cut **4** strips of **25 x 6** flat for guide-strips for tool-pin. File smooth radius on one corner to provide for easy entry of pin.

***Turn over the side member**

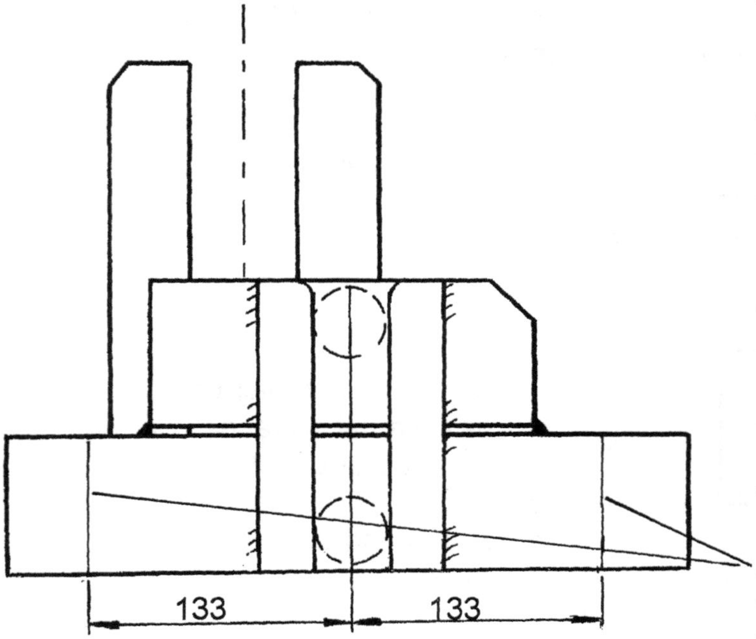

**3.2:** Use 2 pieces of pin material to obtain correct spacing between guides. Tool-pin must be a free sliding fit between guides.

Clamp guide-strips in position - **ensure** they are equidistant either side of marked centre-line.

**3.3:** Weld strips to side-plate. **Do not** weld to inside edge of strip where pin slides.

**3.4:** Carefully measure and mark with scriber) centre-lines for outer bottom roller supports.

**Step 4: Cut and prepare cross members of frame**

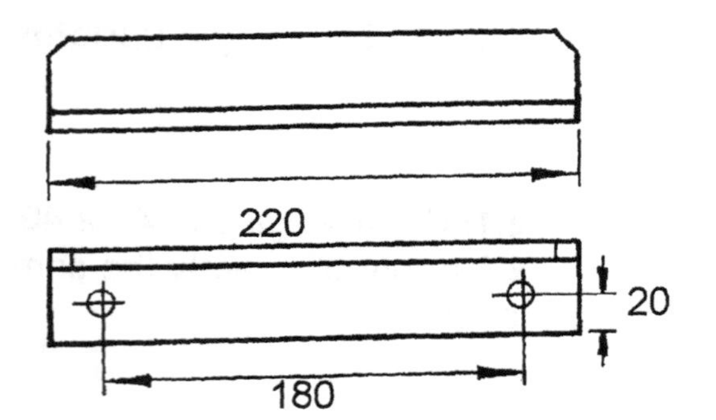


**4.1:** Cut **2** lengths of **40 x 40 x 6** angle for end members.

Drill **2** holes in this face for **M10** anchor bolts.

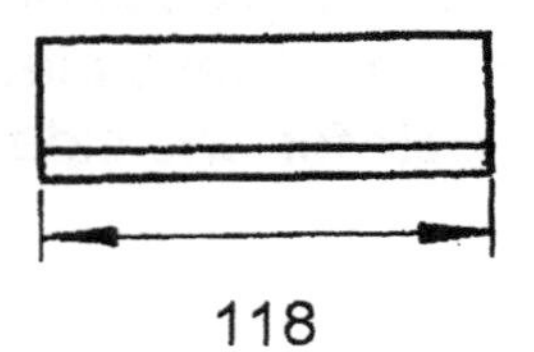


**4.2:** Cut **6** lengths of **40 x 40 x 6** angle for bottom roller supports.

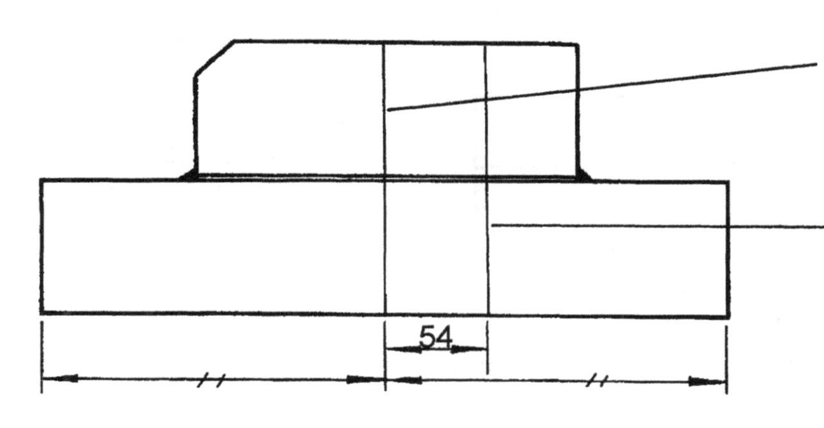


**2.2:** Carefully measure and mark (with a scriber) the centre-lines of each side plate on **both sides** of the plate.

Carefully measure and mark a second line which will be the centre-line between the two angle uprights.

***Use a square to make sure the two lines are square to the base and parallel.**

**2.3:** Mark positions of angle uprights equal distances from centre-line so that a piece of **40 x 40** angle sits neatly in the space as shown.

**2.4:** Clamp 2 angle uprights firmly to side-plates - use a square to ensure uprights are perpendicular to base and check correct positioning with piece of angle.

Tack-weld uprights in position.

**2.5:** Weld in gap between plates **weld alternately from each side to reduce distortion.**

**2.6:** Weld the angle uprights to the side-plates using intermittent welds as shown by ////

**Note:** Since welding is mainly from one side this is likely to cause ends of sideplates to bow upwards. This can be counteracted by clamping the side to produce an opposite bowing effect and welding in this position.

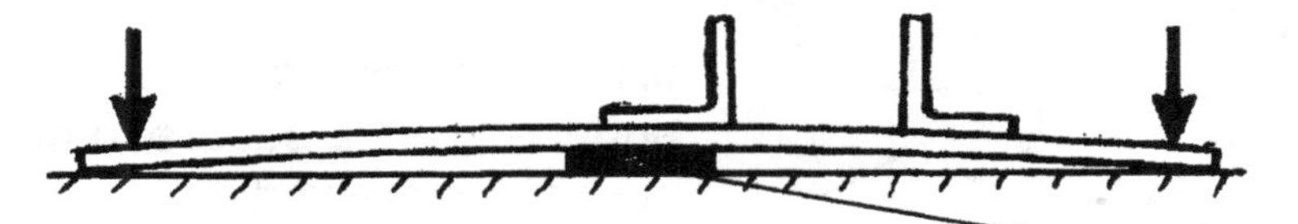

After welding and removing the clamps the sides should hopefully pull flat - if not straighten them.

Clamp side to a rigid beam e.g. channel or 'I' section.

Fit spacer - about 1.5mm thick.

### Step 5: Cut lengths of round bar for pins and rollers

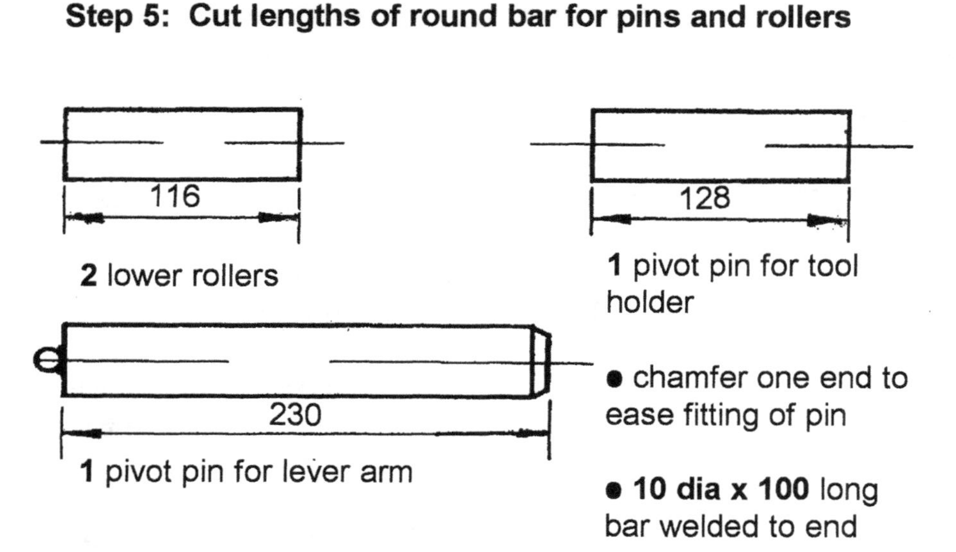


**Note:** preferred material is **40mm** diameter bright mild steel bar (BMS).

**32mm** and **35mm** dia bar may also be used with small changes to lever arm (see later).

### Step 6: Assemble 2 sides of frame

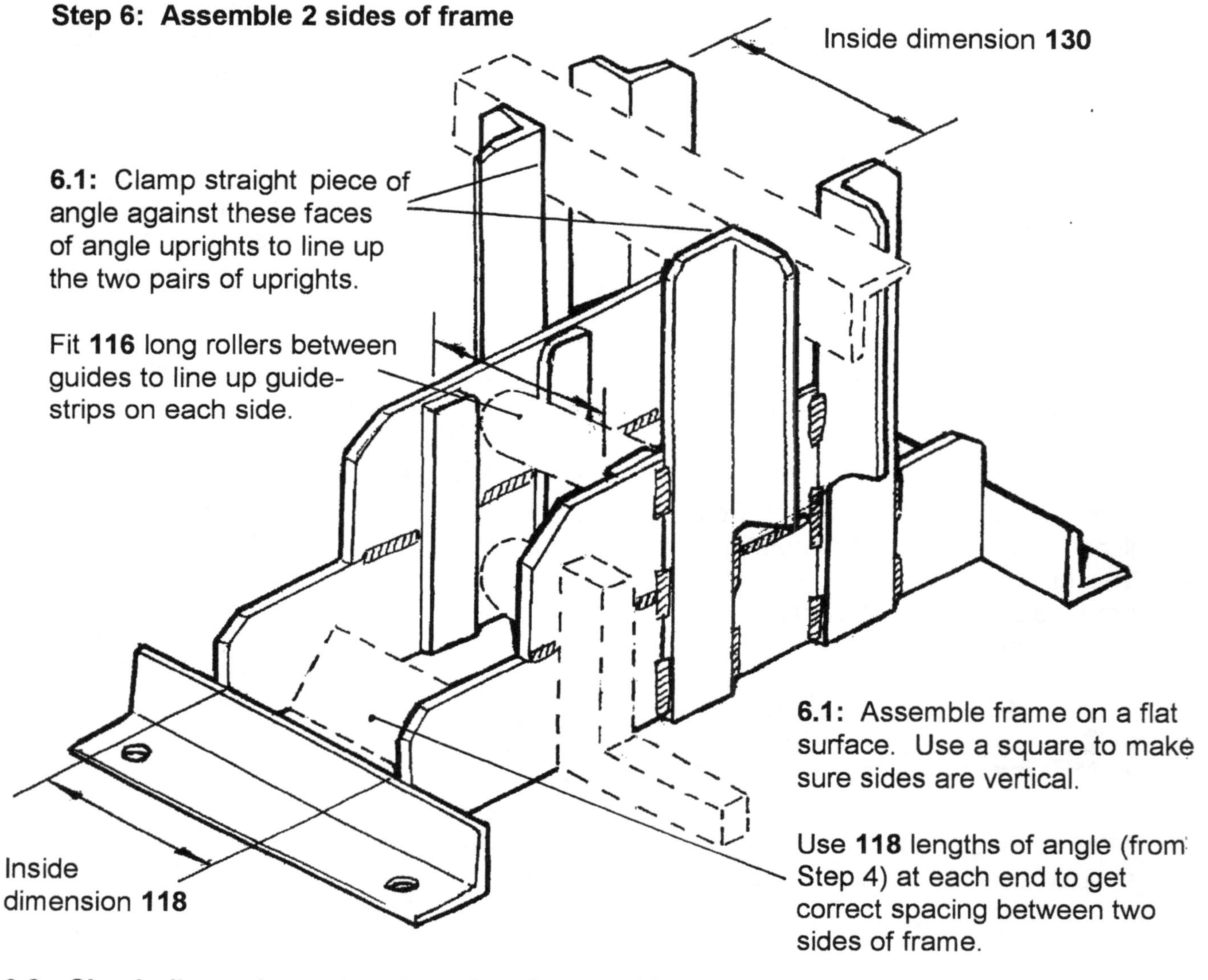


**6.2: Check dimensions at each end and top and bottom.** When correct weld up frame as shown.

**Step 7: Construct housing for the lever-arm pivot pin**

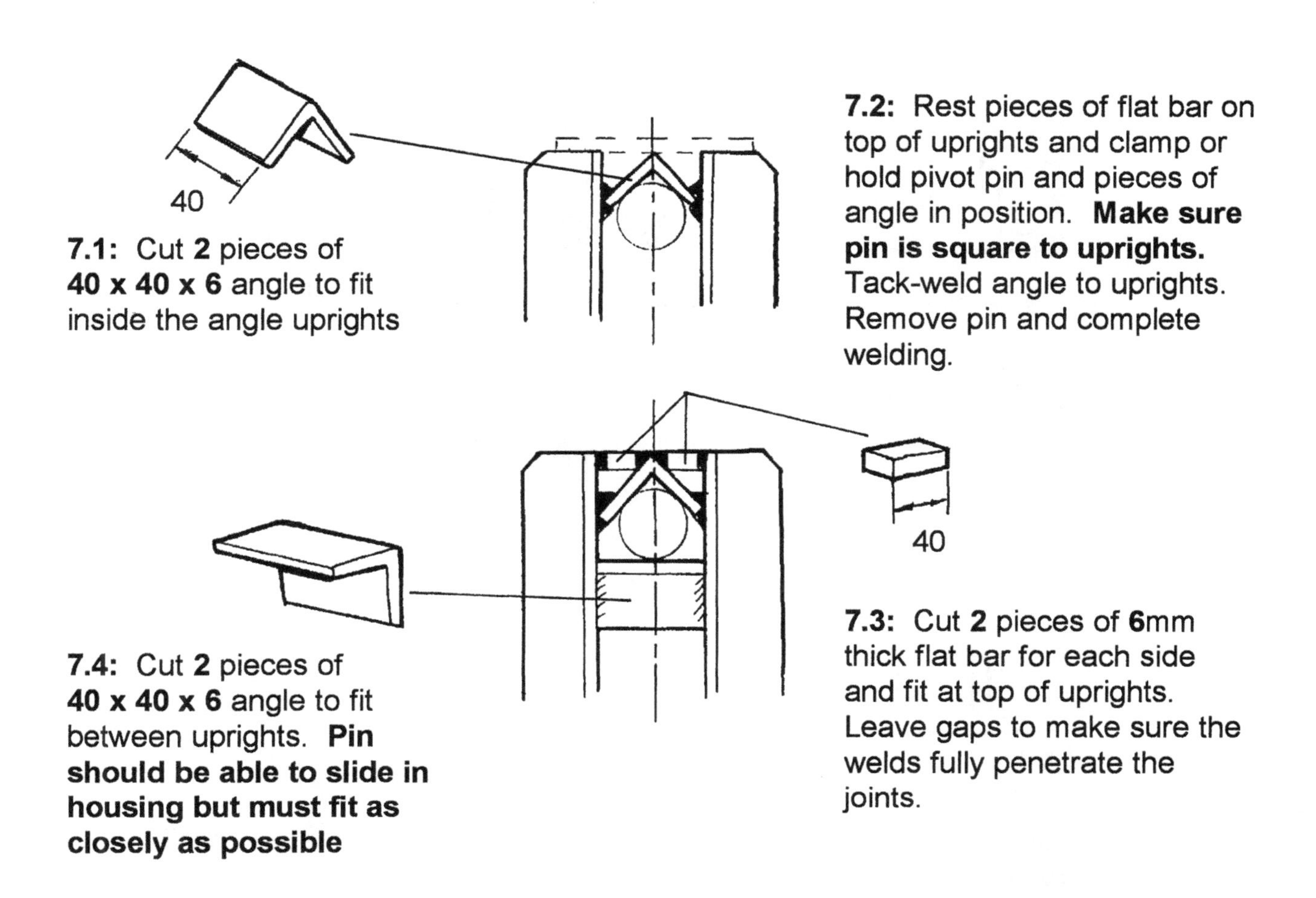


**7.1:** Cut **2** pieces of **40 x 40 x 6** angle to fit inside the angle uprights

**7.2:** Rest pieces of flat bar on top of uprights and clamp or hold pivot pin and pieces of angle in position. **Make sure pin is square to uprights.** Tack-weld angle to uprights. Remove pin and complete welding.

**7.3:** Cut **2** pieces of **6**mm thick flat bar for each side and fit at top of uprights. Leave gaps to make sure the welds fully penetrate the joints.

**7.4:** Cut **2** pieces of **40 x 40 x 6** angle to fit between uprights. **Pin should be able to slide in housing but must fit as closely as possible**

**Step 8: Fit pieces of angle to support lower rollers**

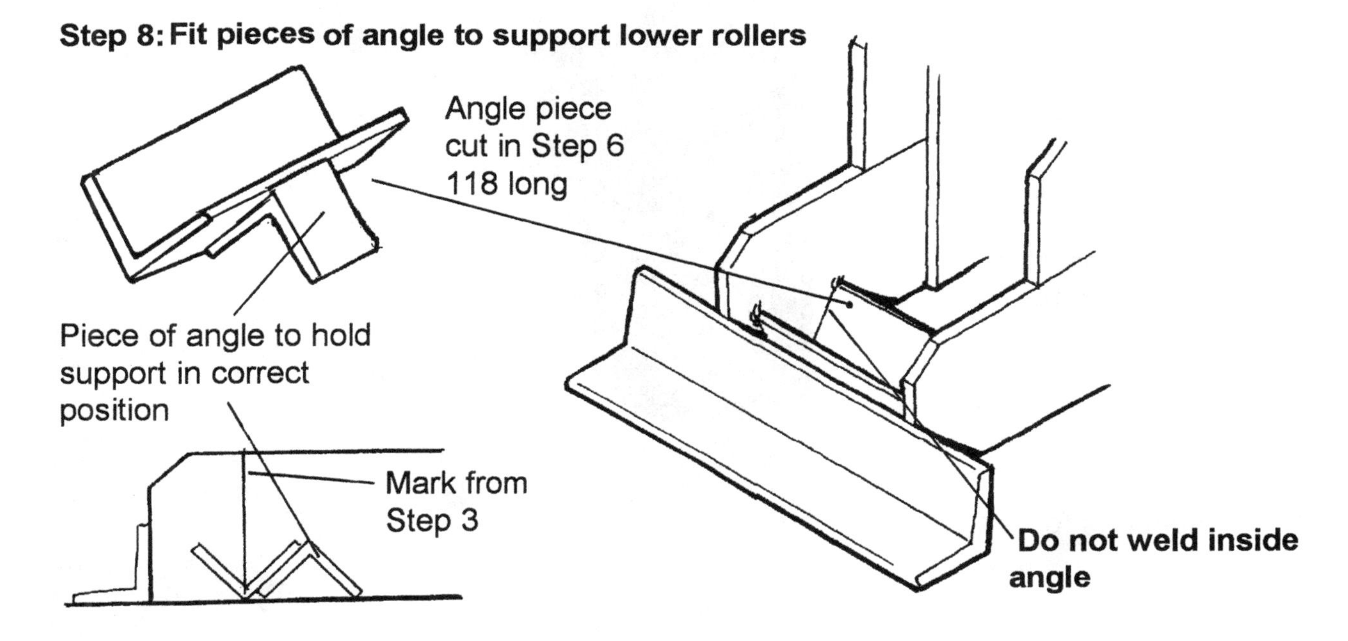


**8.1:** Fit supports for outer rollers inside each end of frame and line up with marks (see Step 3). **Check supports are square to side-plates and same distance apart at each end.** Tack weld supports in position at top corners.

**8.2:** Fit pieces of angle for inner roller supports. Tack-weld them in position

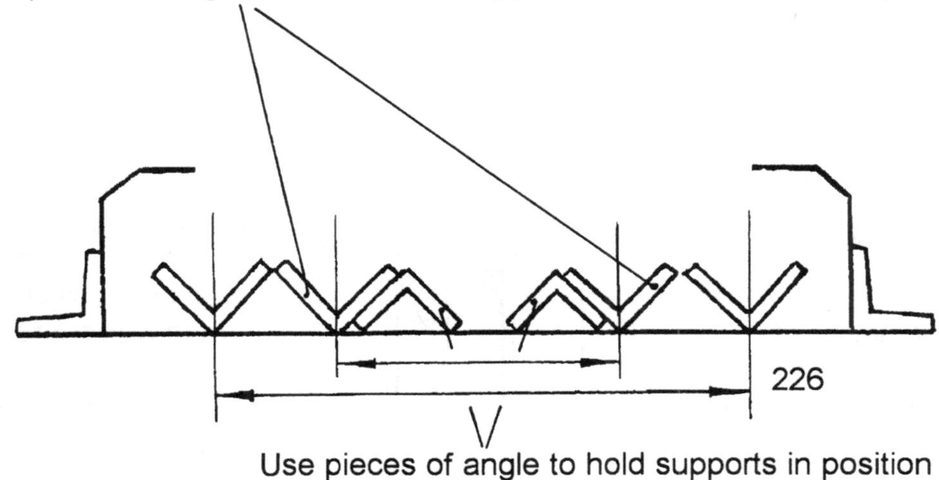


**8.3:** Turn frame on side and weld roller supports along each edge of angle from the underside

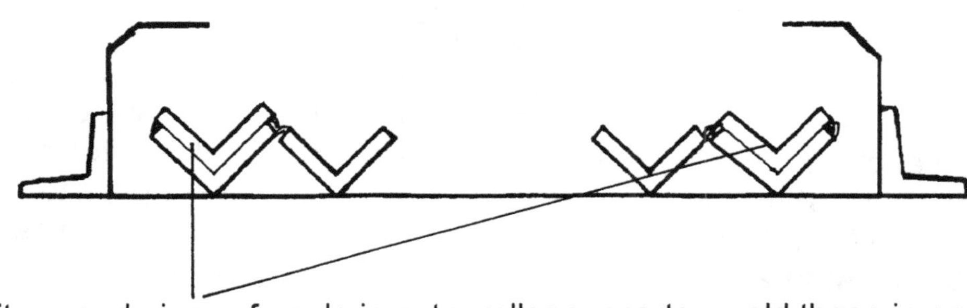

**8.4:** Fit second piece of angle in outer roller supports - weld these in position. **Do not weld inside angle where pin sits.**

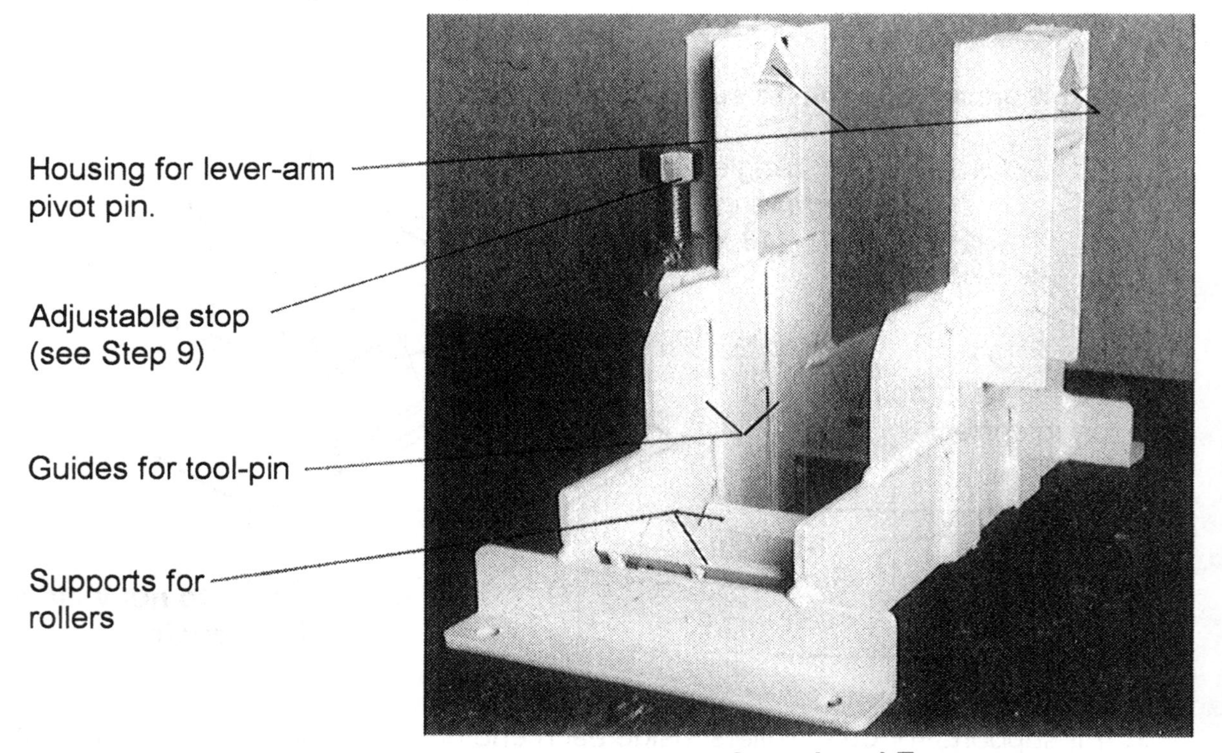


**Completed Frame**

## Step 9: Fit the adjustable stop

**9.1:** Cut a 40mm length of 25 x 6 flat bar

**9.2:** Weld on M20 nut to one end. The nut must be very firmly attached

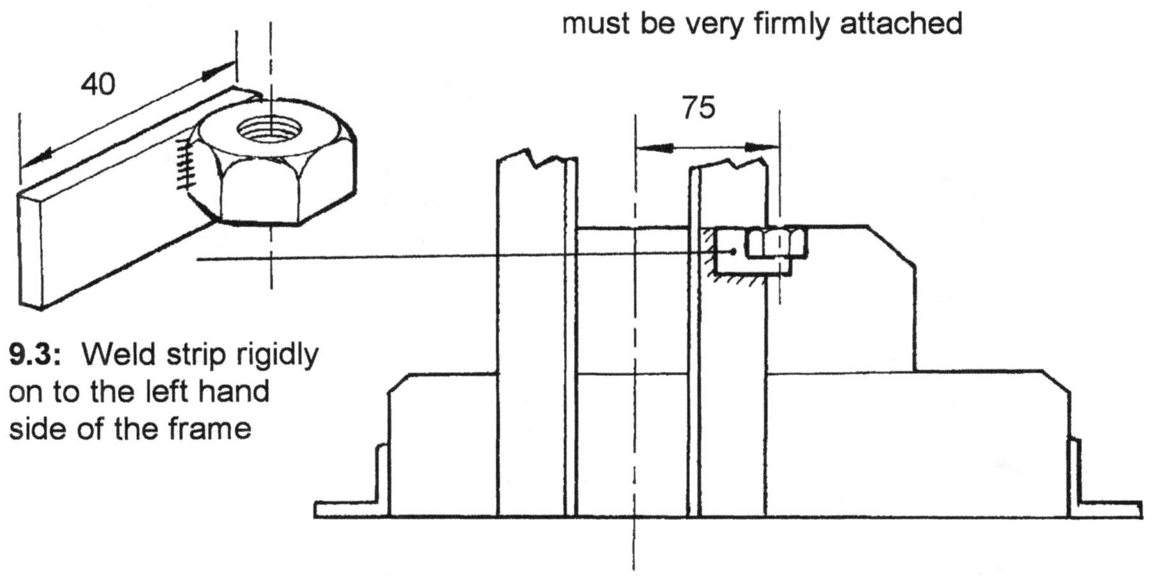


**9.3:** Weld strip rigidly on to the left hand side of the frame

**9.4:** Cut a 150mm length of M20 threaded rod. Weld an M20 nut on one end

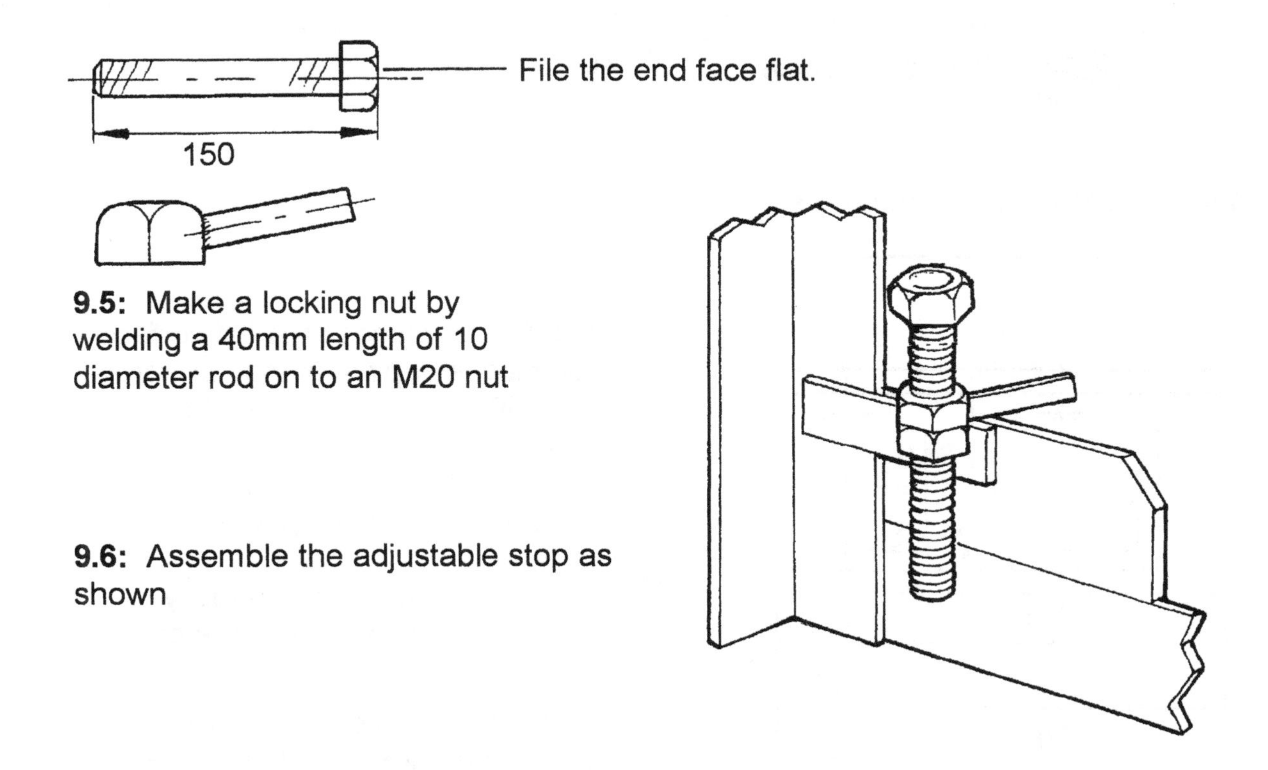


**9.5:** Make a locking nut by welding a 40mm length of 10 diameter rod on to an M20 nut

**9.6:** Assemble the adjustable stop as shown

## 2. LEVER ARM

### Step 1: Cut material for sides

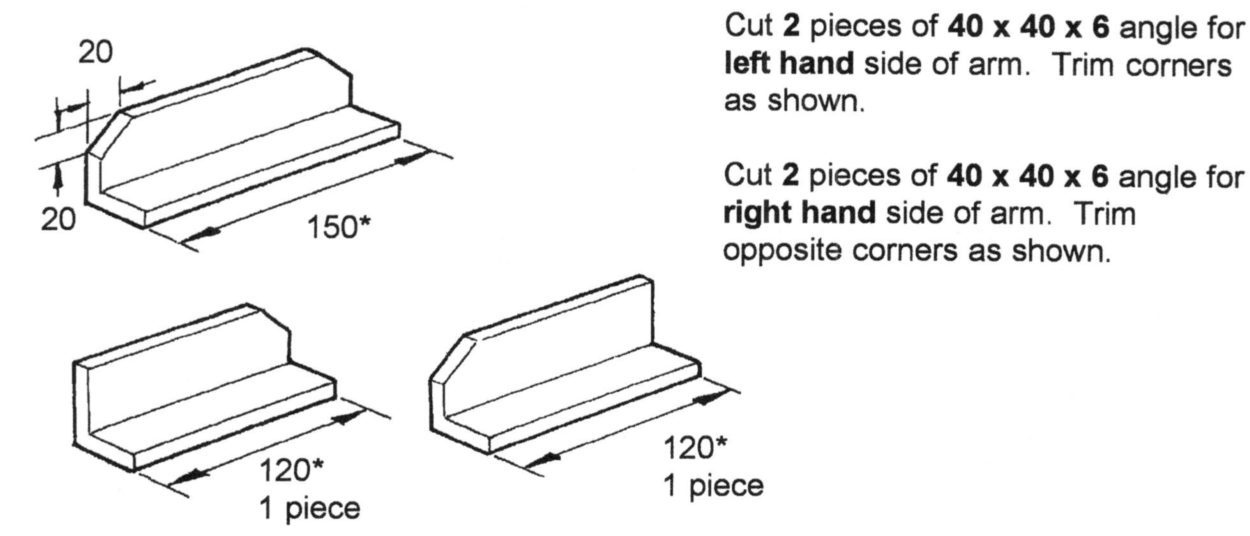


Cut **2** pieces of **40 x 40 x 6** angle for **left hand** side of arm. Trim corners as shown.

Cut **2** pieces of **40 x 40 x 6** angle for **right hand** side of arm. Trim opposite corners as shown.

***Note:** that these lengths are for **38**mm diameter pins. If **40**mm diameter pins are used, **add 2mm** to these dimensions.

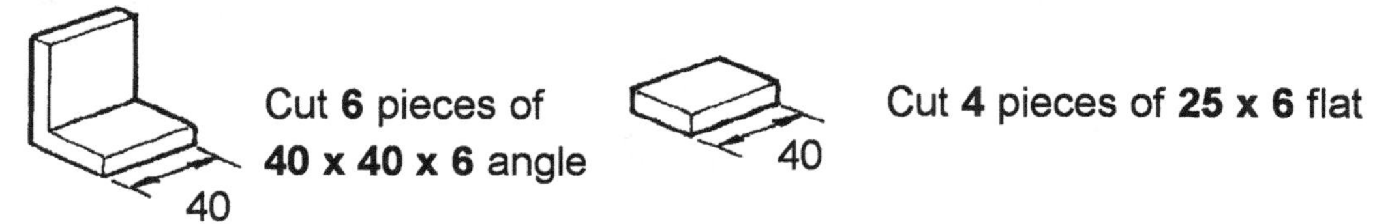


Cut **6** pieces of **40 x 40 x 6** angle

Cut **4** pieces of **25 x 6** flat

### Step 2: Construct right hand side of lever arm

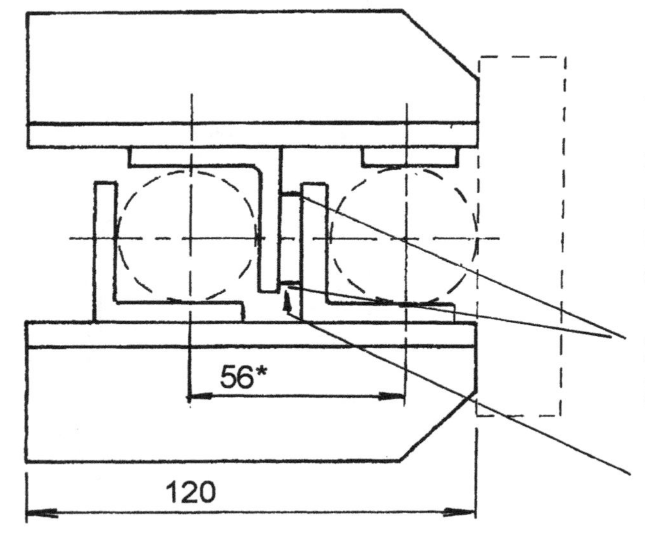


* If **40**mm diameter pins are used this distance will be **58**mm.

**2.1:** Set up pieces on a flat surface. Use pieces of pin material (preferably 38 or 40mm diameter but 32mm may also be used) to line up pieces of angle and to get correct spacing. Use straight edge to line up ends.

**2.2:** When pieces are correctly positioned tack-weld at these points.

** If **32**mm diameter pins are used fit **2** pieces of **25 x 6** flat here to keep correct spacing of **56**mm.

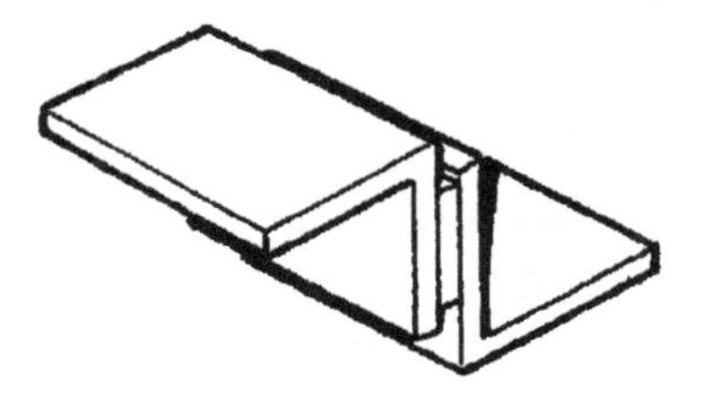

**2.3:** Take out the parts that have been tack-welded together and complete welding along the joints.

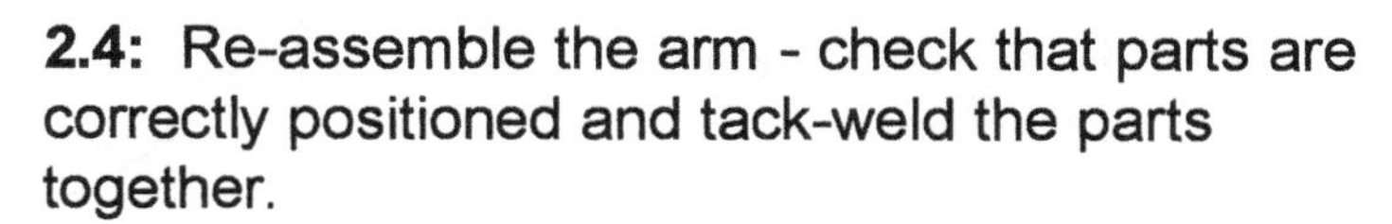

**2.4:** Re-assemble the arm - check that parts are correctly positioned and tack-weld the parts together.

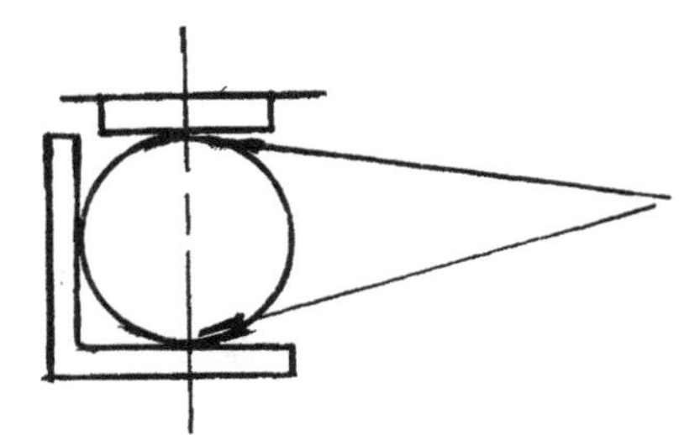

**Note:** that the pins should be a tight fit.

File shallow grooves where pins seat so that pins rotate and slide freely but are still a close fit.

**2.5:** Remove the pins and weld up around all joints as fully as possible.

**Step 3: Make left-hand side of lever arm**

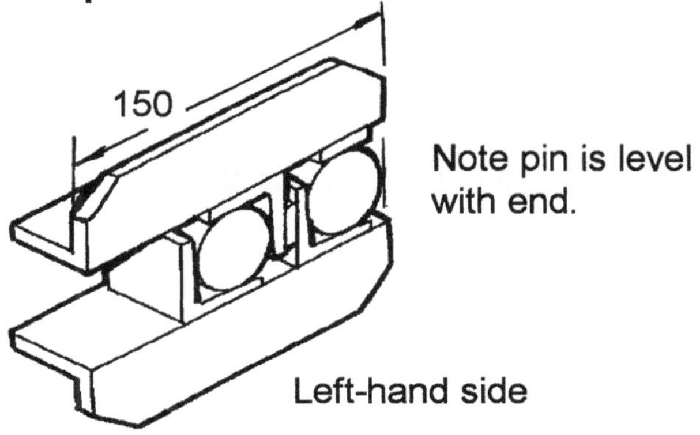


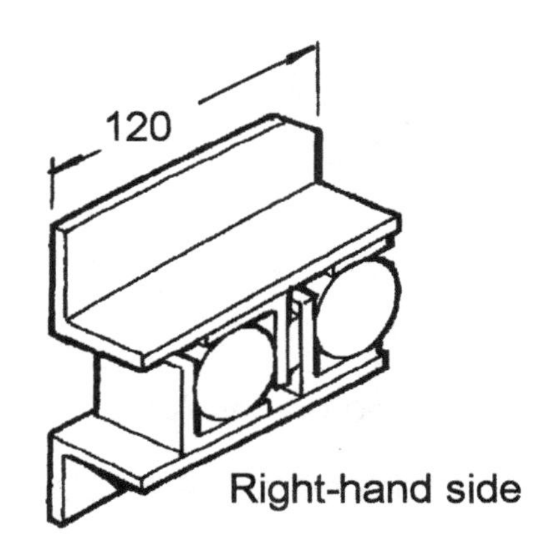


Set up parts in similar way to Step 2 to make a matching side as shown above.

**Step 4: Cut the cross members to join the 2 sides**

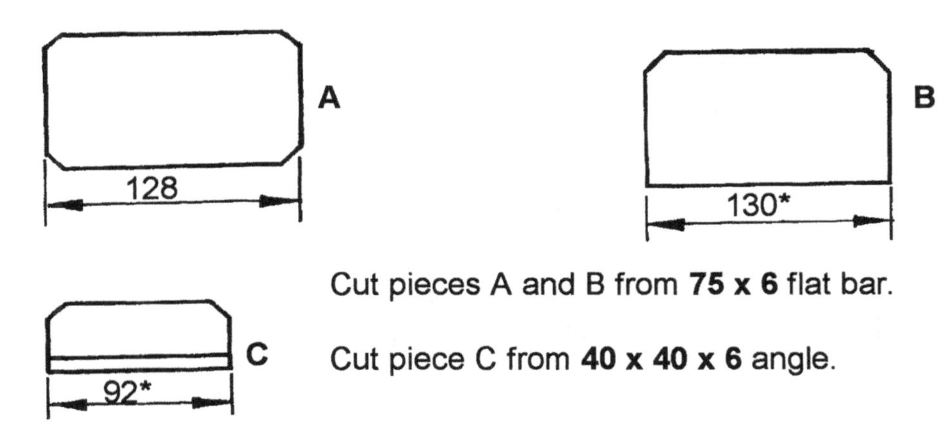


Cut pieces A and B from **75 x 6** flat bar.

Cut piece C from **40 x 40 x 6** angle.

***Note:** that these dimensions will be **2mm** larger if **40mm** diameter pins are used.

## Step 5: Assemble the lever arm

**5.1:** Set up the two sides using the pivot pins to line up the sides.

**5.2: Note** that outside width of arm must be **128mm** to fit in main frame with clearance of 1mm each side.

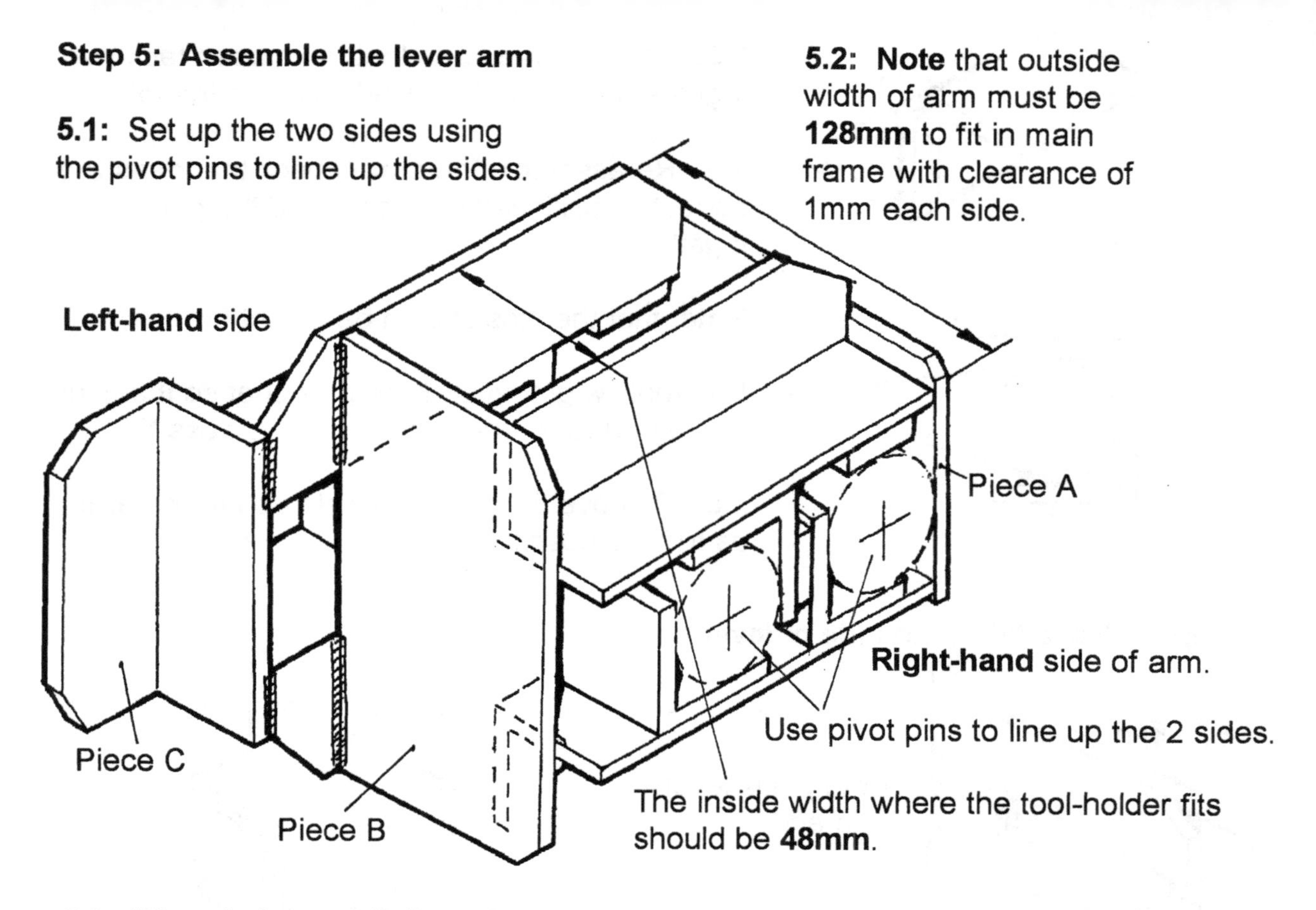


**5.3:** Fit end pieces A,B,C, and tack-weld them to sides.

**5.4:** Check set-up is satisfactory:

- pins should be able to slide and rotate freely but should fit as closely as possible in housing;
- outside width should be **128mm at each end**.

**5.5:** Weld up around all joints with pins in position.

## Step 6: Make up socket for hand lever

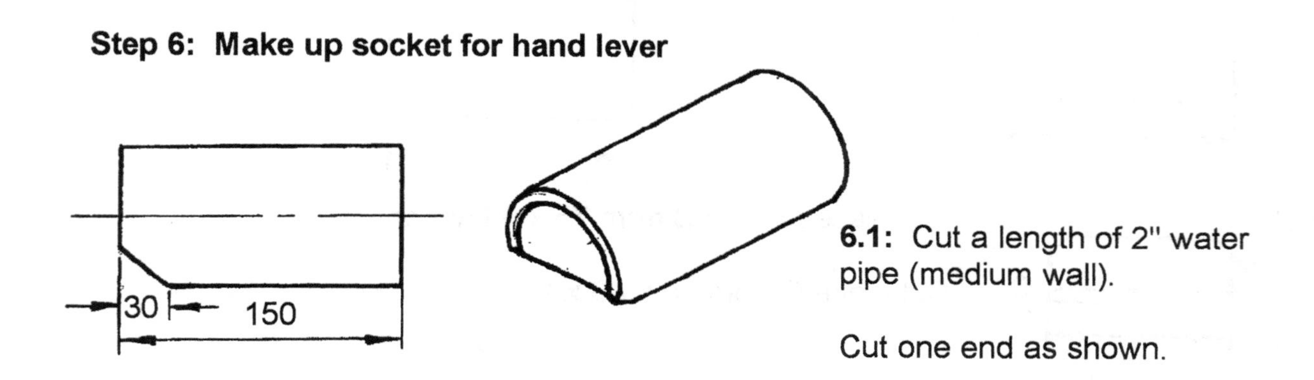


**6.1:** Cut a length of 2" water pipe (medium wall).

Cut one end as shown.

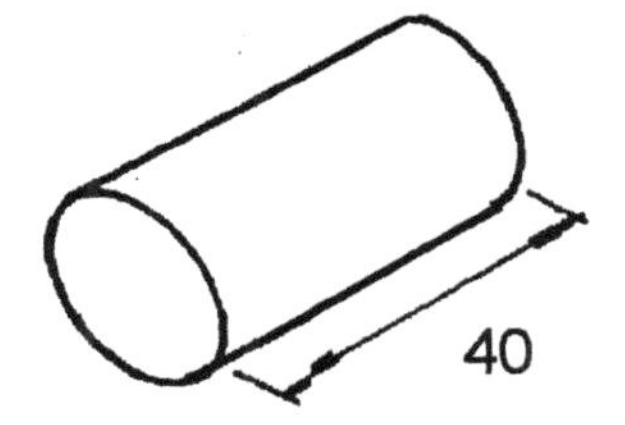


**1** piece **20** or **25** dia
round bar or **3/4**" pipe.

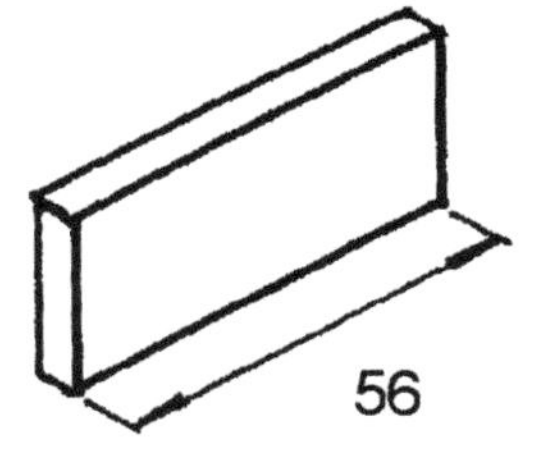


**1** piece **25 x 6**
flat bar.

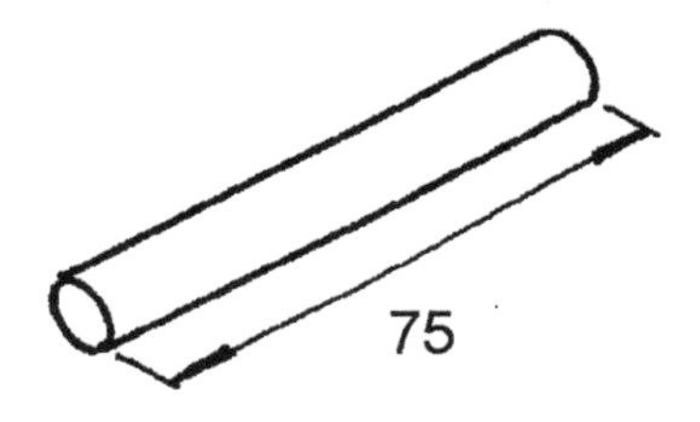


**2** pieces **10** dia
round bar.

**6.2:** Weld the piece of **25 x 6** flat across the end of the pipe socket as shown.

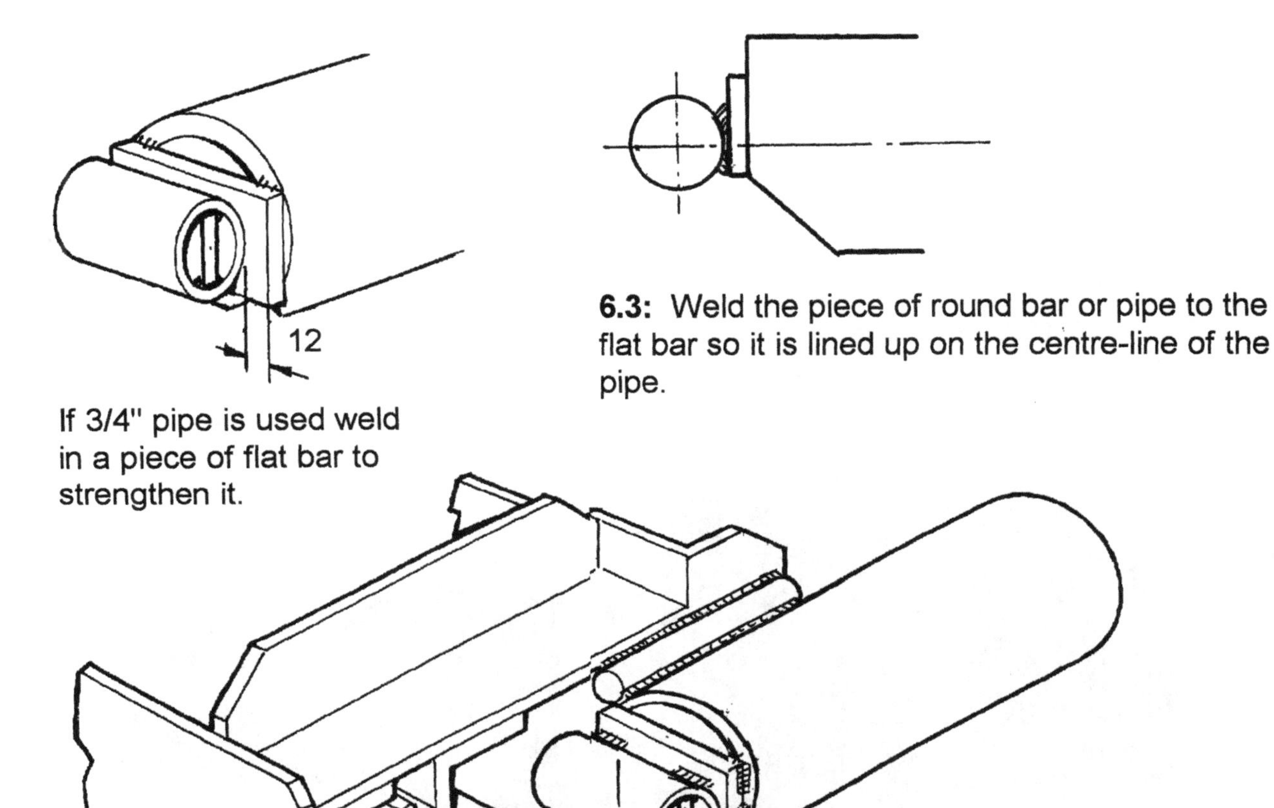


If 3/4" pipe is used weld in a piece of flat bar to strengthen it.

**6.3:** Weld the piece of round bar or pipe to the flat bar so it is lined up on the centre-line of the pipe.

**6.4:** Fit the lever arm in the main frame using the pivot pin and line up the pipe socket along the left-hand side of the arm so that the **stop sits centrally on the adjustable stop.**

Weld the pipe to the side of the arm.

Reinforce the weld by welding in the 10 diameter rods.

**Completed Lever Arm**

## 3. TOOL HOLDER

### Step 1: Cut lengths of 40 x 40 x 6 angle

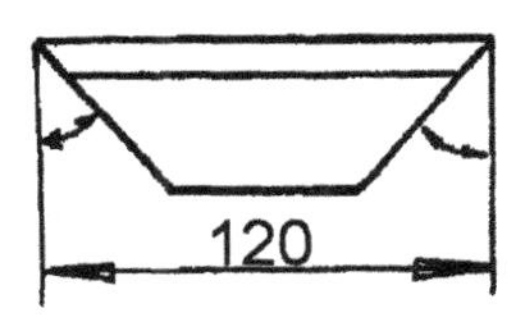


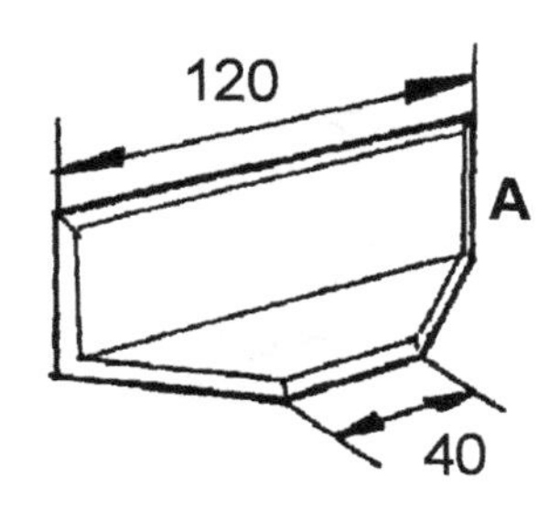


Cut **2** pieces of angle as shown.

Cut **1** piece of angle

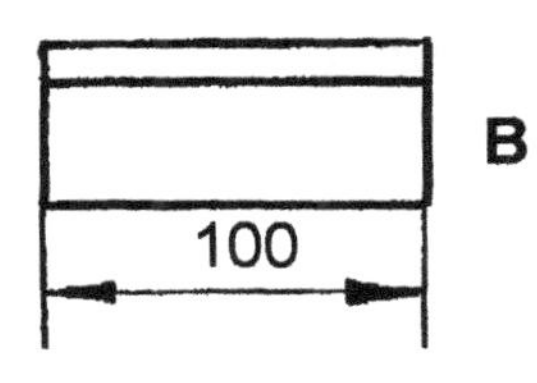


Cut **1** piece of angle

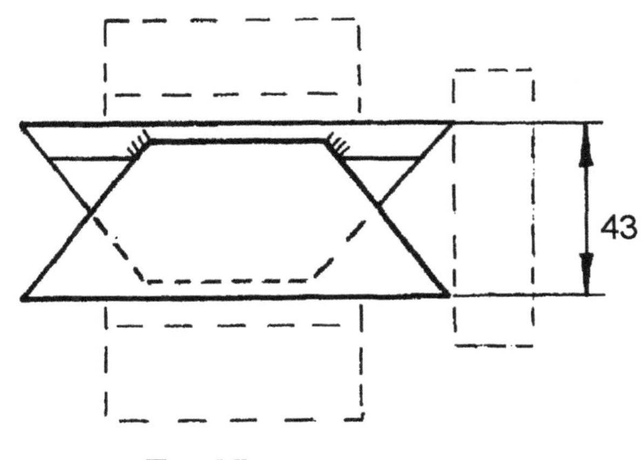


**Top View**

### Step 2: Assemble the tool holder

**2.1:** Fit 2 pieces of angle A together to form a box section **43mm** wide.

Line up ends with square. Use pieces of angle to hold members A in correct position.

Tack-weld top and bottom.

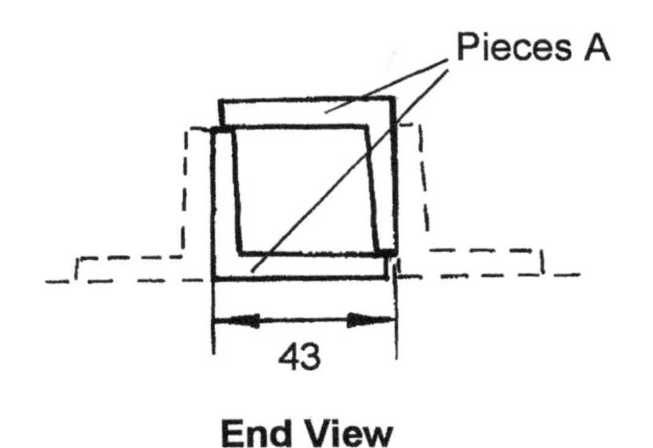


**End View**

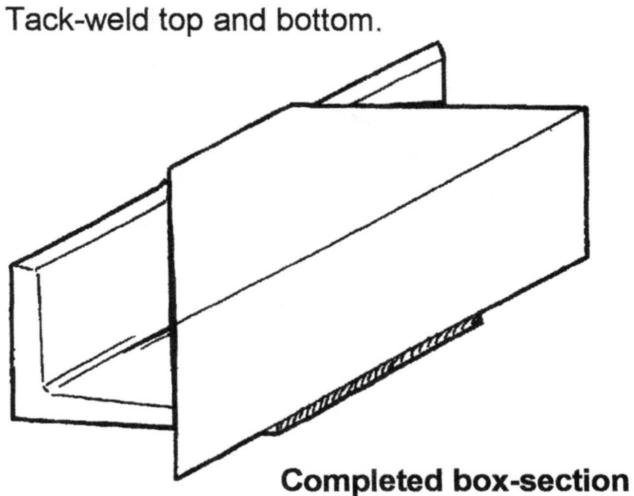

**Completed box-section**

**2.2:** Fit pieces of angle B and C into the ends of the box member.

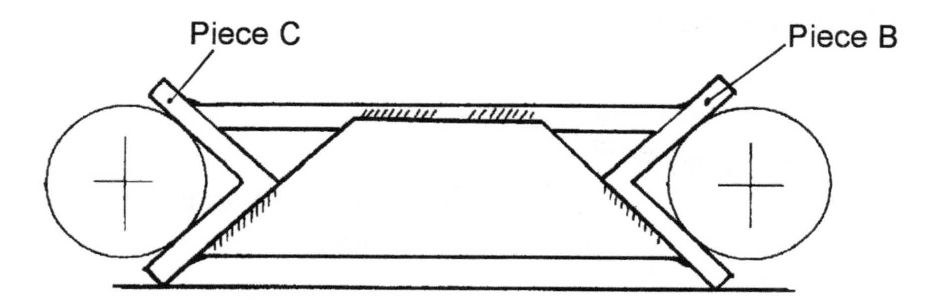


Fit pins into the angle at each end and hold them in position so that they are parallel. **Check carefully that pins are parallel** i.e. ends are same distance apart and also same height above a flat surface - and then weld pieces B and C to box member.

**2.3:** Weld on pieces of flat bar to hold pins in position.

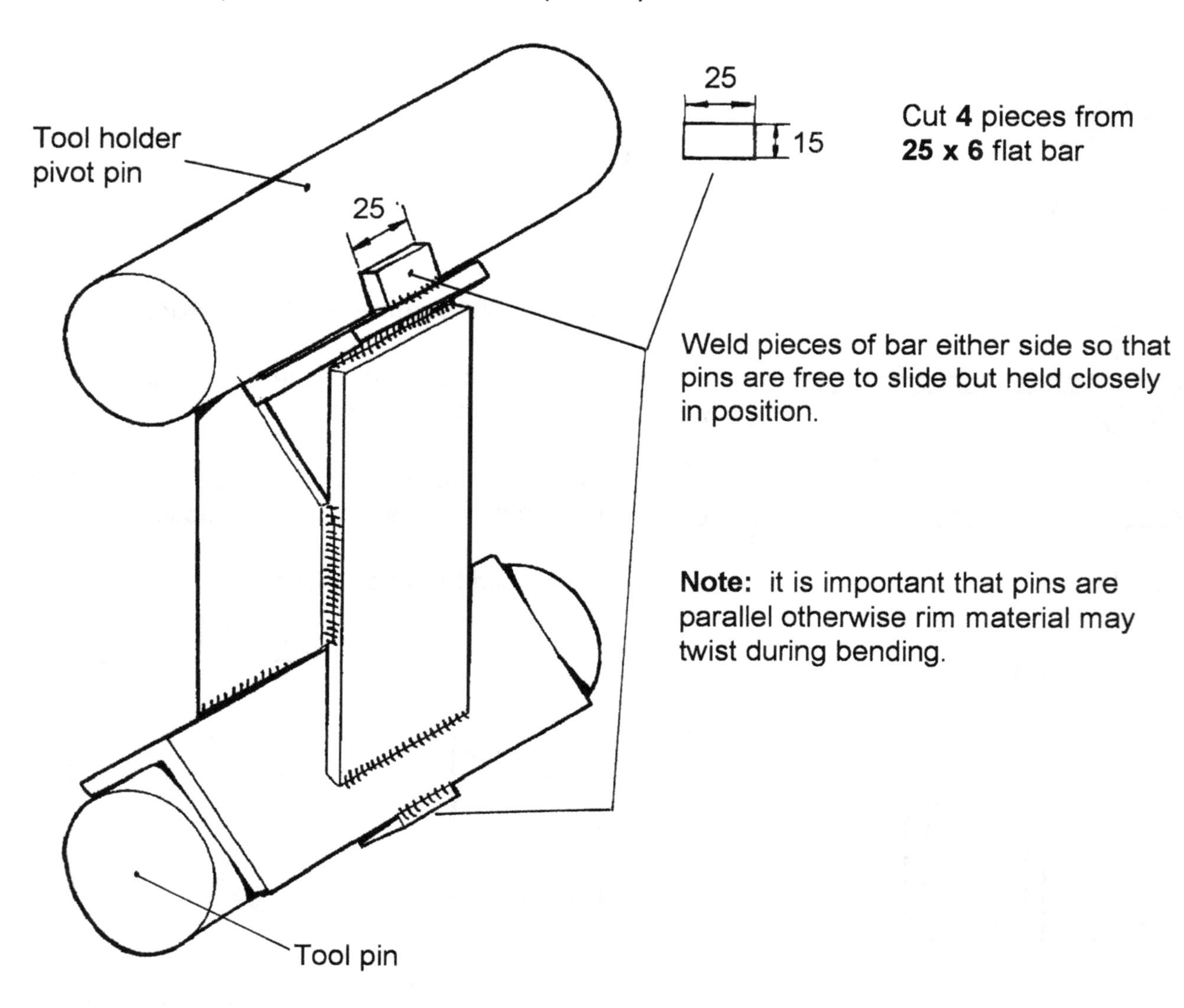


Cut **4** pieces from **25 x 6** flat bar

Weld pieces of bar either side so that pins are free to slide but held closely in position.

**Note:** it is important that pins are parallel otherwise rim material may twist during bending.

## 4. TOOLS (see Drawing 107/1/103)

### 4.1: Construct 'V' tool for bending angle and round sections

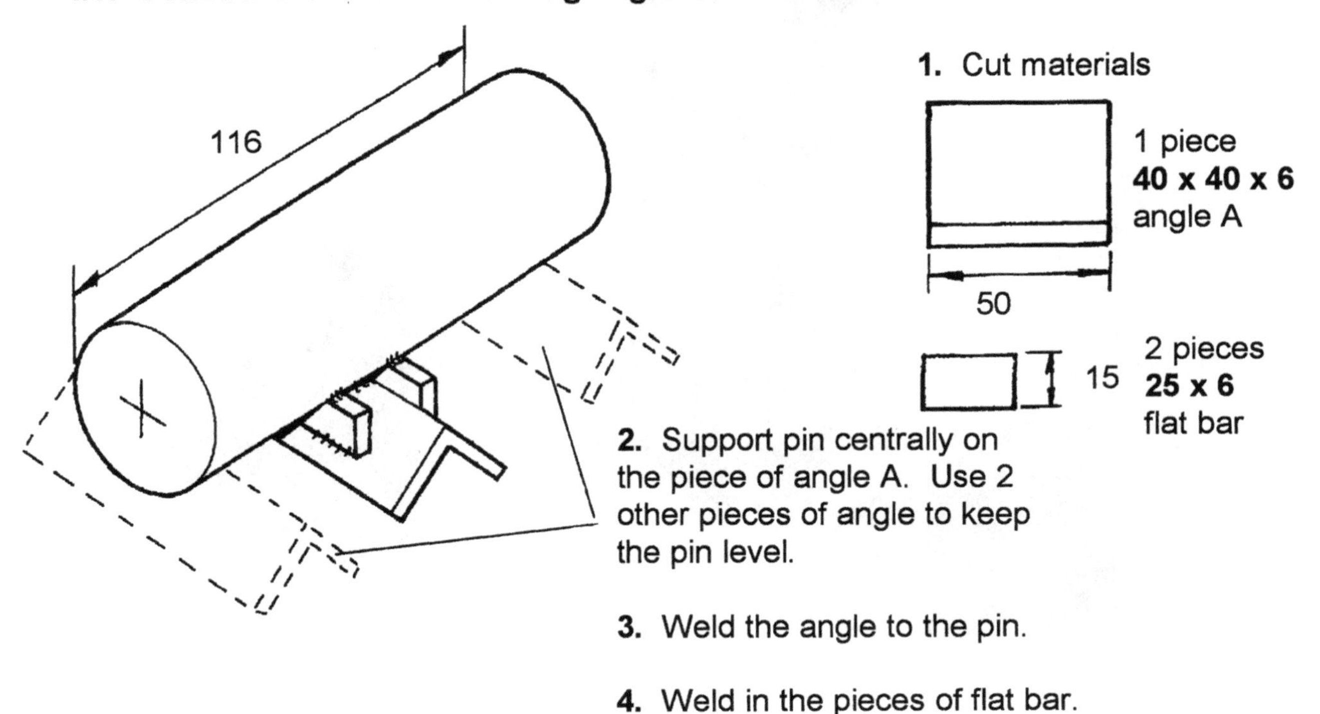


1. Cut materials

1 piece **40 x 40 x 6** angle A

2 pieces **25 x 6** flat bar

2. Support pin centrally on the piece of angle A. Use 2 other pieces of angle to keep the pin level.

3. Weld the angle to the pin.

4. Weld in the pieces of flat bar.

### 4.2: Construct 2 point bending tool for bending flat bar

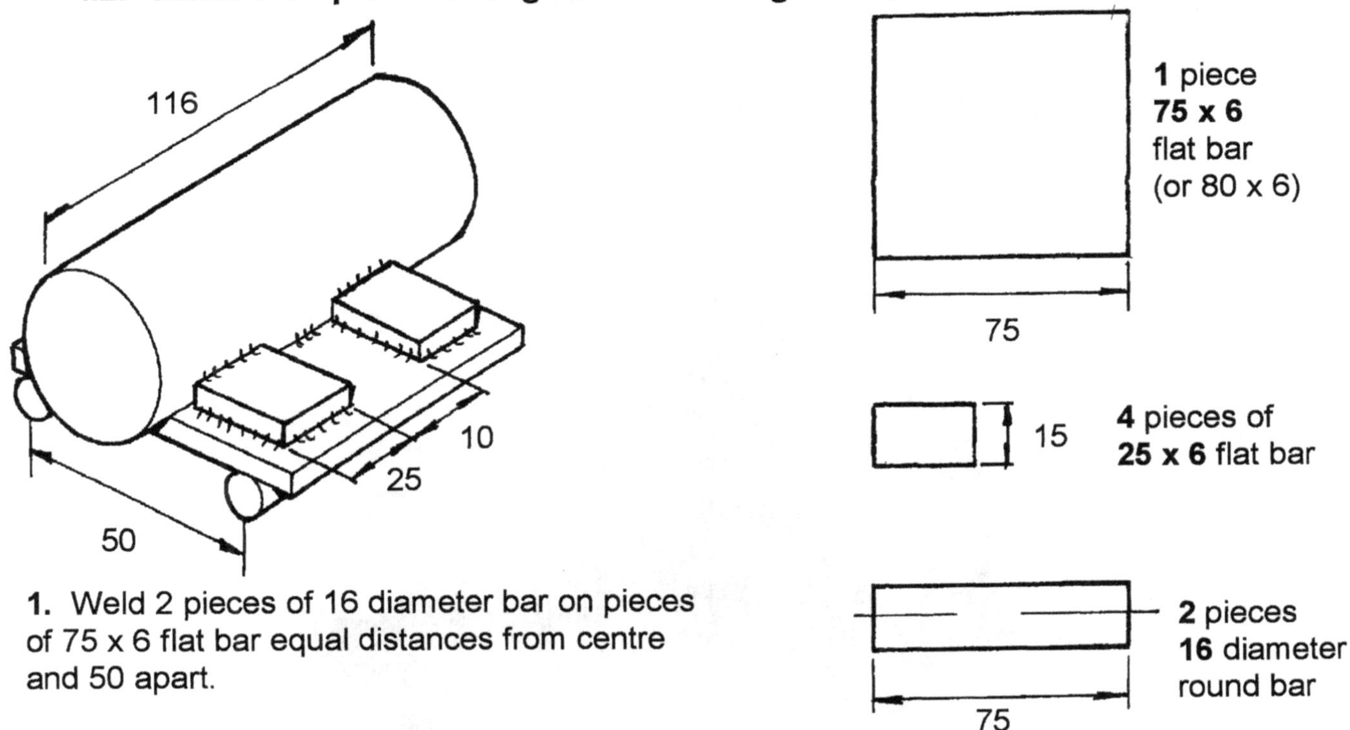


**1** piece **75 x 6** flat bar (or 80 x 6)

**4** pieces of **25 x 6** flat bar

**2** pieces **16** diameter round bar

1. Weld 2 pieces of 16 diameter bar on pieces of 75 x 6 flat bar equal distances from centre and 50 apart.

2. Fit tool pin **centrally** on other side of plate **mid-way** between the 2 pieces of 16 rod.

3. **Check** tool pin is parallel to pieces of 16 rod and weld pin to plate.

4. Weld the 4 pieces of 25 x 6 flat to the plate and pin to reinforce the joint.

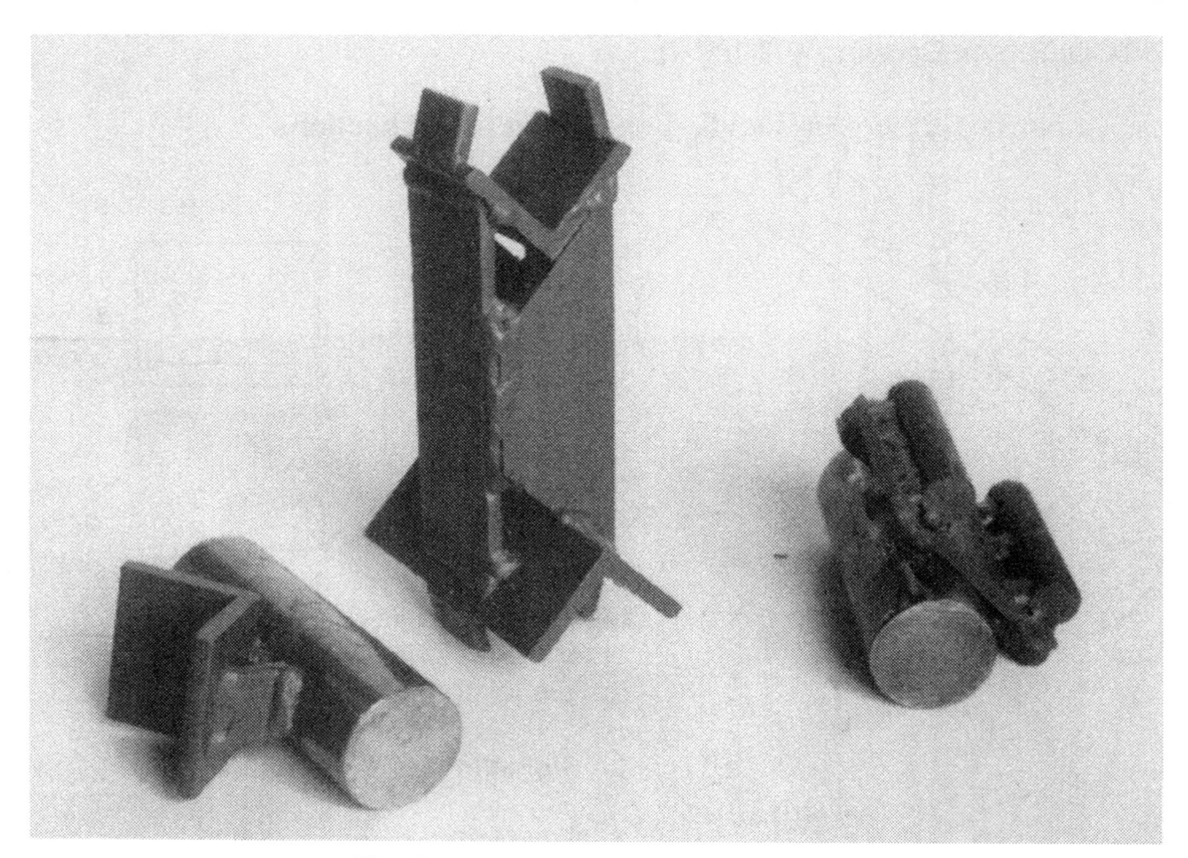

**Tool Holder and Forming Tools**

**Assembly of Lever Arm and Tool Holder**

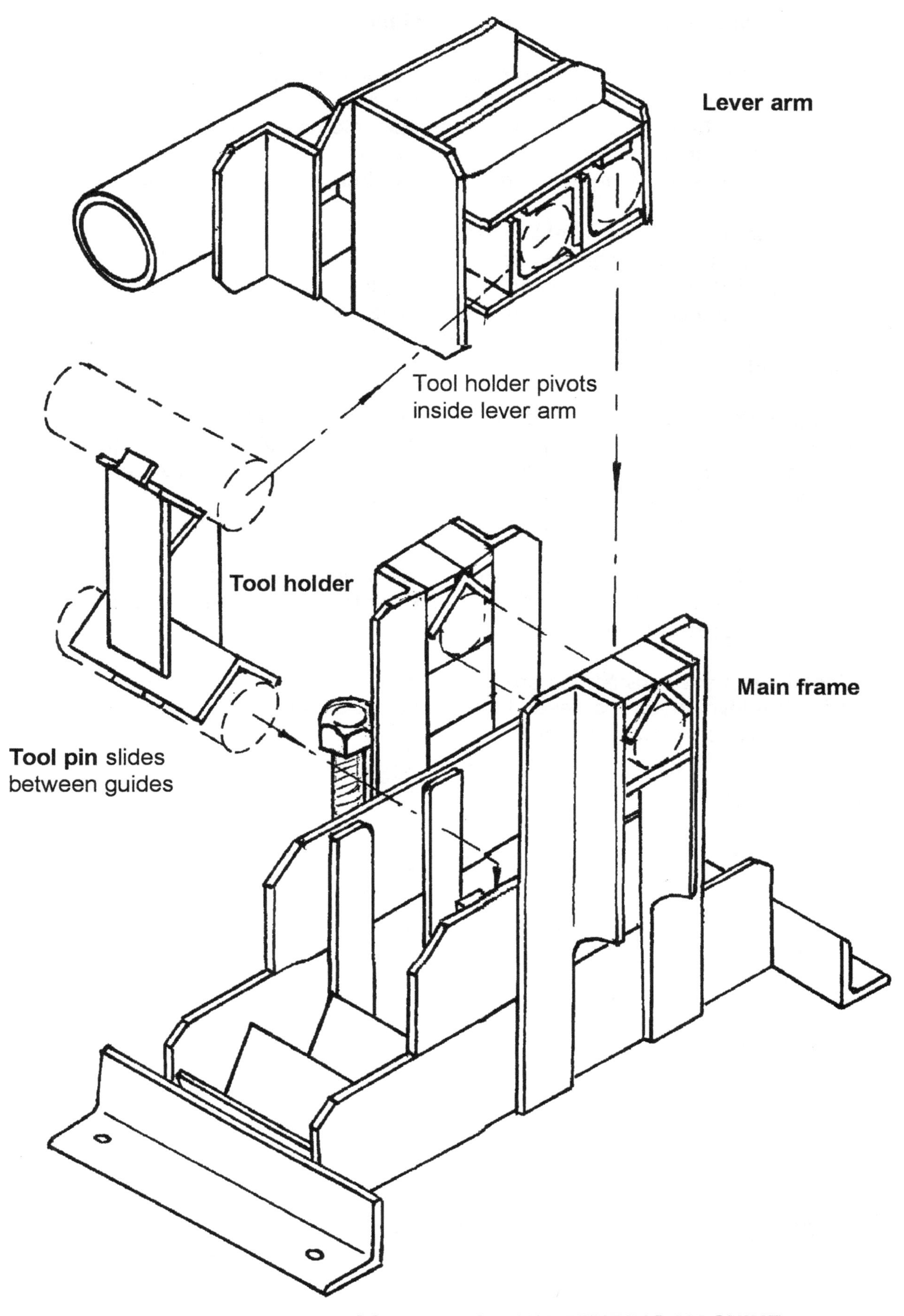


ASSEMBLY OF RIM-BENDING MACHINE

## Alternative Design of Rim-Bending Machine

The rim-bending machine which has been described has been designed for construction in a workshop which does not have a drilling machine for drilling large size holes. The pivot pins are therefore fitted into housings formed by angle section rather than into holes. The need to cut, position and weld in these pieces of angle adds some time and complication to the construction process.

If a drilling machine is available which can drill holes of at least 32mm diameter, a simpler and neater design can be constructed. This is shown in the adjacent drawing. The main change is that the pivot pins are fitted into machined holes in flat bar. The angle uprights of the main frame are replaced by 65 x 12 or 80 x 12 flat bar and the construction of the lever arm is simplified by using 65 x 12 flat bar for the sides.

The preferred size of flat bar for the sides of the main frame is 65 x 12. Wider materials, 75 or 80 wide, may be used but the heights of the side uprights and position of the pivot hole will need to be adjusted so that their overall height is the same (i.e. for 80 wide bar the heights of the side uprights will be 90 - 15 = 75 and 200 - 15 = 185 and the height of the pivot hole centre 170 - 15 = 155).

10mm thick bar may be used if 12mm is not available but this is the minimum thickness to be used. If 10 or 12mm thick bar is not available then weld together two thicknesses of 6mm thick flat bar.

The steps in the construction are the same as set out in the previous section and the same level of care is needed to cut and set up parts accurately to ensure the bender will perform satisfactorily.

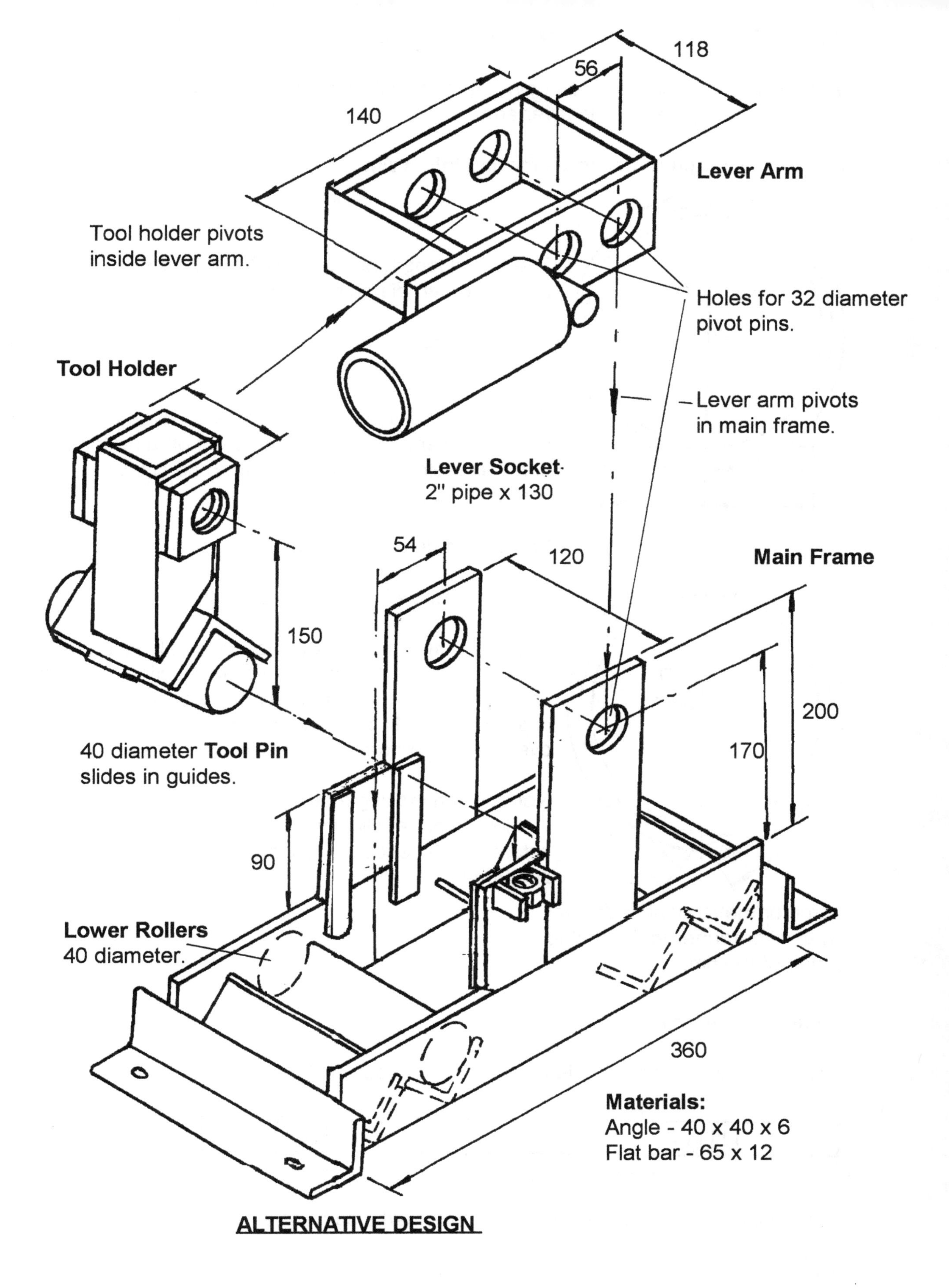


ALTERNATIVE DESIGN

## 2.3: OPERATION OF THE RIM-BENDING MACHINE

1. **Mount the machine** on a heavy, sturdy bench or stand which will not topple over when force is applied to the lever arm.

2. **Cut the length of bar to be formed into a rim**

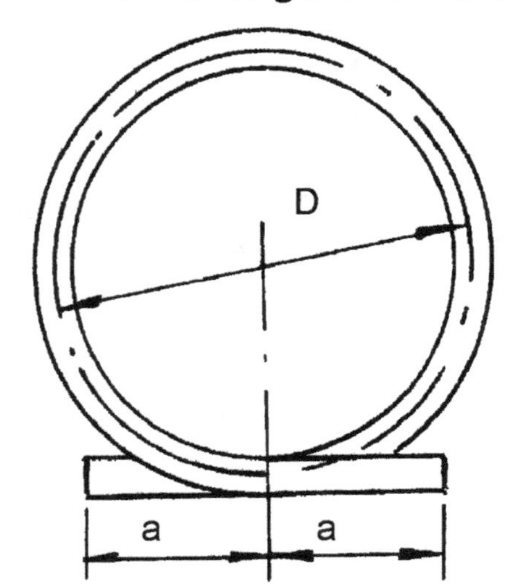


The actual length of the rim = **π.D = 3.14 x D.**
**D** is the **mean** diameter of the rim i.e. outside diameter - thickness.
However when the rim is bent, straight parts are left at the two ends.

To bend the full length of the rim, allowances **'a'** must be left at each end and these will be cut off after the rim is bent. Therefore the length of bar to be cut will be:
**3.14 x D + 2a**

For lower rollers in the **outer** position **2a = 350mm**.

For lower rollers in the **inner** position **2a = 200mm.**

When the rim is bent the two ends **'a'** will have to overlap. This is not possible with some stiffer sections, generally flat bar, and in these cases the rim bar should be cut to the exact length, **3.14 x D**, and the straight ends hammered over an anvil or former to the required curvature.

**Note:** when making a new type of rim it is best to cut the first length of bar slightly too long and then to estimate the actual length needed by subtracting the lengths that have to be cut off.

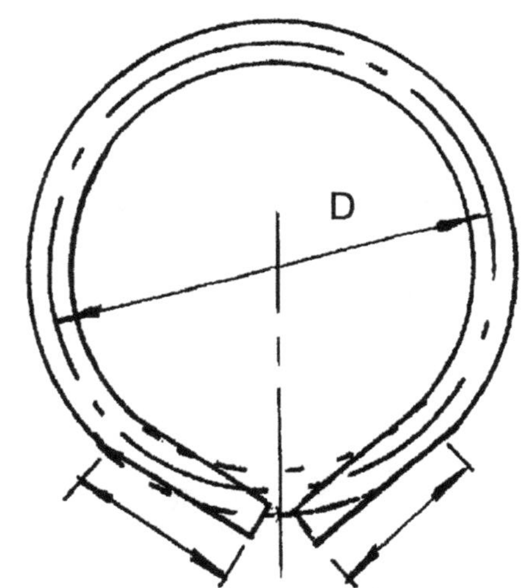


Hammer the straight ends over an anvil or former to the dotted shape of the rim.

**3. Mark out the bar**

**3.1:** If allowances are included mark off **'a'** at each end.

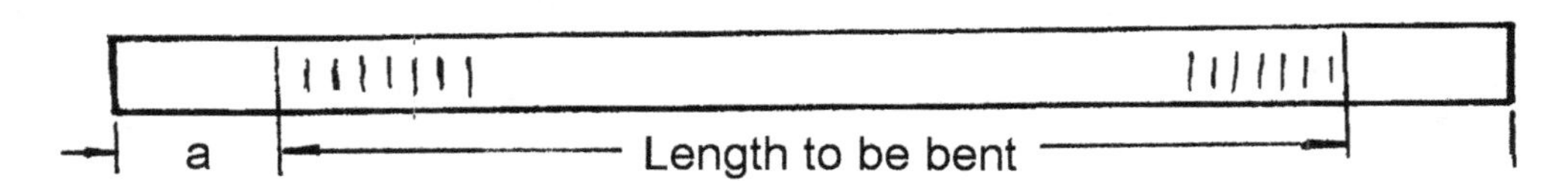


**3.2:** Mark the length to be bent with chalk marks as follows:

- **25mm** steps for the lower rollers in the **inner** position
- **50mm** steps for the lower rollers in the **outer** position.

**Note:** If there is any tendency for flats to be formed on the bar during bending reduce the size of the steps.

**4. Set up the machine**

**4.1:** Fit the correct forming tool for the section to be bent into the tool holder.

**4.2:** Fit the lower rollers into the required inner or outer position depending on the size of rim to be bent.

**5. Adjust the machine for the size of rim to be produced** (see drawing)

**5.1:** Zero the machine, lay a **straight** length of rim bar on the lower rollers and rest the forming tool on the bar with a **slight** downward pressure on the lever arm. Screw the adjusting screw up until it **just touches** the stop.

**5.2:** Find the number of turns the adjusting screw has to be screwed down for the rim size that is to be bent. Average settings are given in the instructions for making standard wheels and the charts in Figure 2.2 should also be used as a guide. Further notes on setting the adjusting screw are given below.

**6. Bend the bar**

**6.1:** Adjust the bar on the lower rollers so that the leading end is just resting on the furthest roller.

**6.2:** Fit an extension bar in the lever socket and bend the rim bar down until the stop touches the adjusting screw. **Make sure the stop touches the adjusting screw but do not exert excessive force on the stop.**

**6.3:** Lift the lever arm and push the rim bar through to the next chalk mark. Bend the bar again.

**6.4:** Keep pushing the bar through the bender in the steps shown by the chalk marks and bending the bar at each step. **Do not miss any steps and make sure the stop touches the adjusting screw each time the bar is bent. Keep the bar along the centre of the rollers and the bent part upright - do not let it twist sideways.**

**6.5:** If extra lengths have been added at the ends of the bar, towards the end of bending the two ends of the bar will overlap. Push the leading end to the right-hand side of the bender where the bottoms of the angle uprights have been cut away.

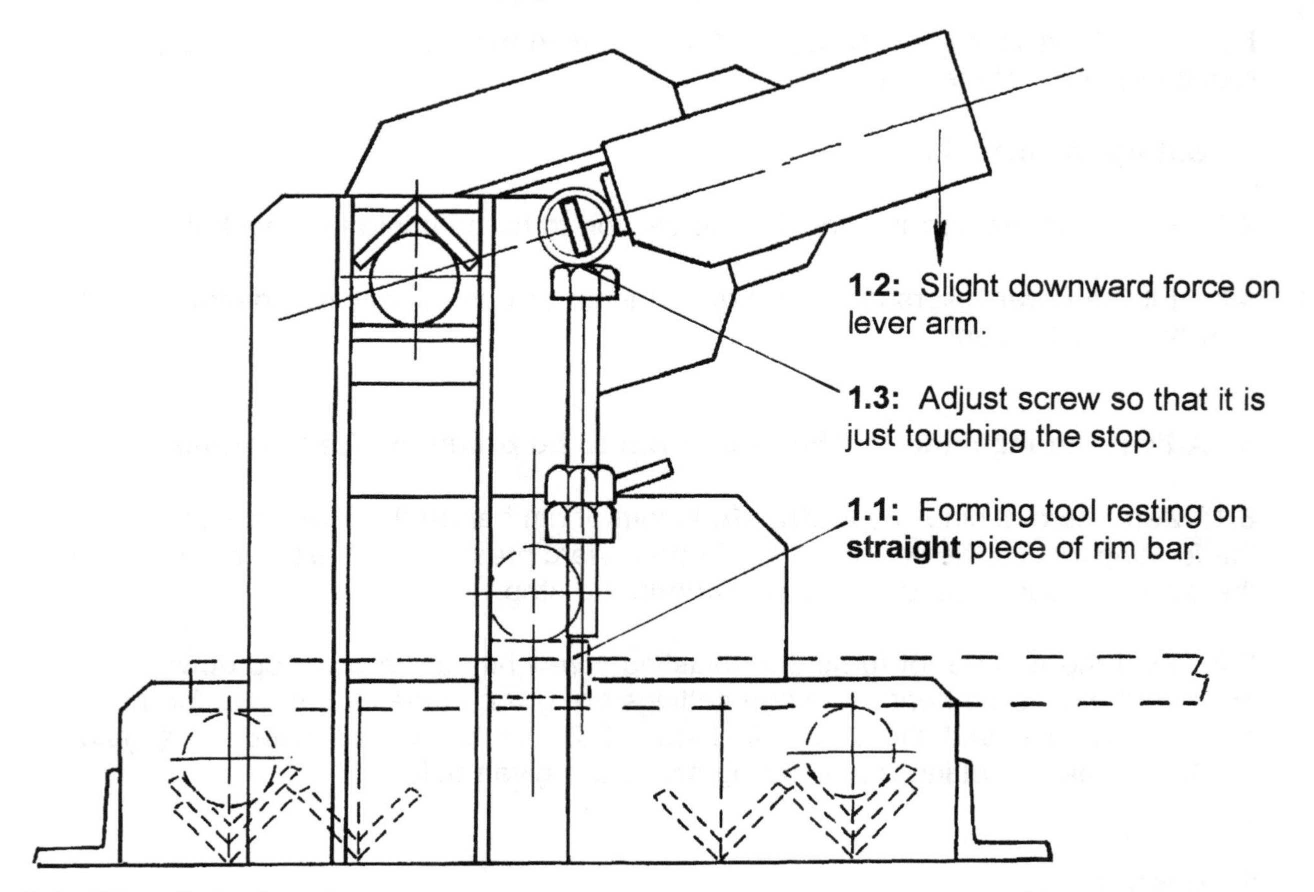


**5.1: "Zero" the bender**

**5. 2: Set the adjusting screws to get the required rim diameter**

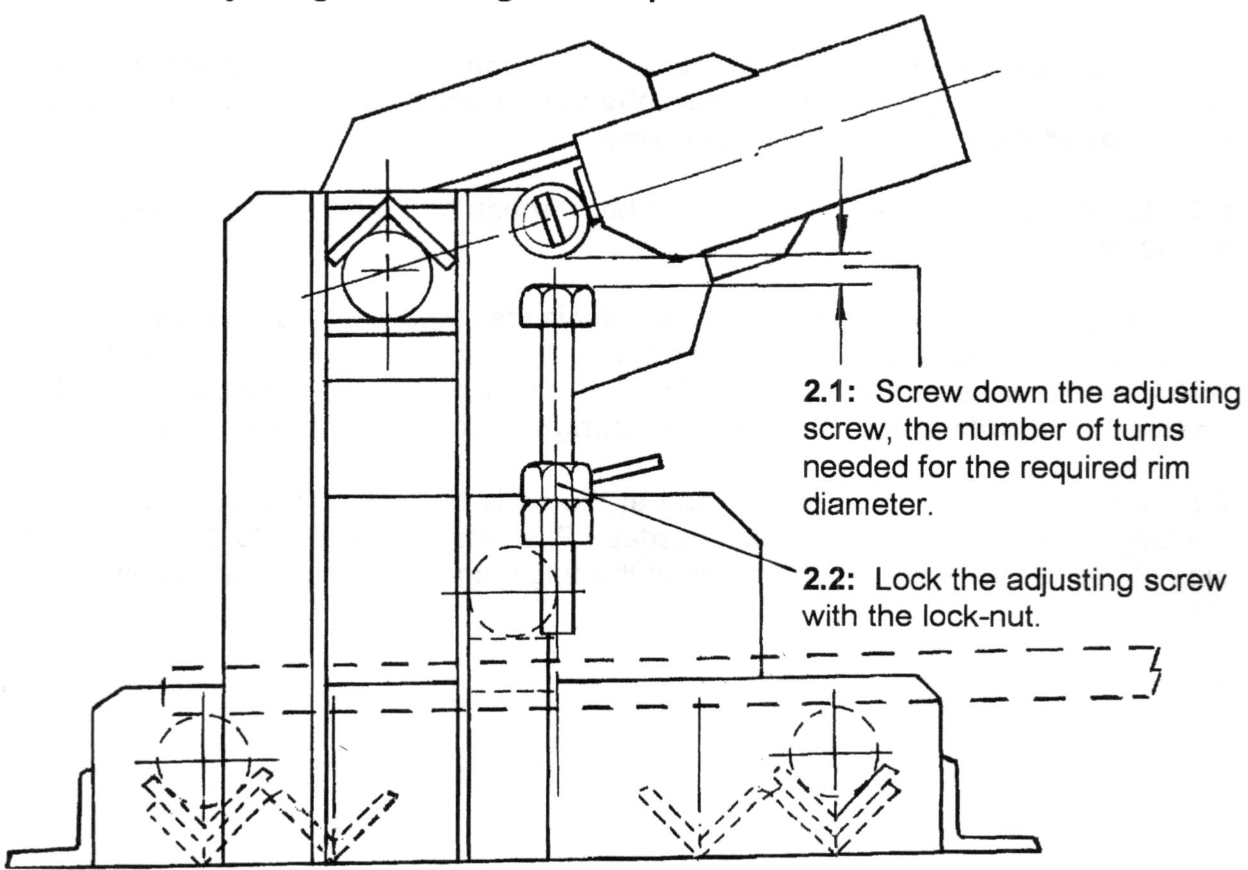

## 7. Check the size of the formed rim

When the rim is fully bent, measure its outside diameter.

**7.1:** If the diameter is within an allowable range of that required (within about 5 to 10mm), the rim is satisfactory and may be removed from the bender.

**7.2:** If the rim is too small, remove it and hammer around its whole length to make it larger. Then put it back and re-bend it. **(Note:** the rim must be bent open around its whole length and not just by pulling the two ends open).

**7.3:** If the rim is too large, adjust the stop position and re-bend the rim (see notes below).

**7.4:** To remove the rim from the bender, remove the pivot pin and lift the lever arm out of the frame.

**8.** The rim is now ready for cutting to length and setting up in the assembly jig. Specific instructions are given in the wheel construction sections.

### Notes on adjusting the machine

When the bar is bent there is some "spring back" which depends on the hardness of the bar material. Softer steel will not spring back as much as harder steel. Therefore softer steel will bend more and form a smaller rim size than harder steel.

The charts and settings given in this manual are average values which can be used as guidelines but some experimentation may be needed to get the exact setting when starting a new size of rim or changing to a new batch of bar material. This is quite straight forward if a step by step approach is taken.

**1.** Choose: **Chart A** for the lower rollers in the **inner** position.
**Chart B** for the lower rollers in the **outer** position.

**2.** Each chart has two lines:

- use **First Bend (lower)** line when bending from **straight**

- use **Re-bend (upper)** line when the rim has already **been bent and has to be re-bent** to correct the size.

**3.** If a setting of the adjustable stop is not given for the rim size to be made choose a setting from the appropriate chart. If the material seems soft screw the adjustable screw down about 1/2 turn less than that given by the chart.

**4.** Bend the rim - even if it does not seem to be coming out at the correct size complete the full rim - **never stop part way through bending a rim** as it will then be almost impossible to get a round rim.

**5.** When the rim is fully formed, measure its outside diameter and write this down and also the number of turns. Note where this point lies on the chart and moving vertically up the chart (i.e. at the same diameter) mark a point about the same distance from the re-bend (upper) line. Then moving parallel to the upper line estimate the number of turns to get the required diameter when re-bending the rim.

**6.** To get the correct initial setting (bending the rim from straight) go back to the lower point and move parallel to the lower line to get the number of turns for the required diameter.

**7. Note:** for **Chart A** (inner position of lower rollers) **1 turn** of adjusting screw changes the rim diameter by **70 to 75mm**.

for **Chart B** (outer position of lower rollers) **1 turn** of adjusting screw changes the rim diameter by **55 to 60mm** (between diameters of 500 and 700mm).

**Example of procedure**

A rim with an outer diameter of 400mm is required.

**1.** Set lower rollers in **inner** position.

**2.** Use **Chart A, lower line**, to find number of turns = **5.**

**3.** Bend rim, measure diameter - this is found to be **430mm**.

**4.** Note this point **(1)** on Chart A and move upwards to point **(2)** same distance above upper line.

**5.** Move parallel to upper line to required diameter of 400mm - point **(3)**. This gives number of turns for re-bend of **7 1/6**.

**6.** To correct initial setting, from point **(1)** move parallel to lower line to diameter of 400mm - point **(4)**. This gives setting for bending of next rim (from straight) as **5 1/3** turns.

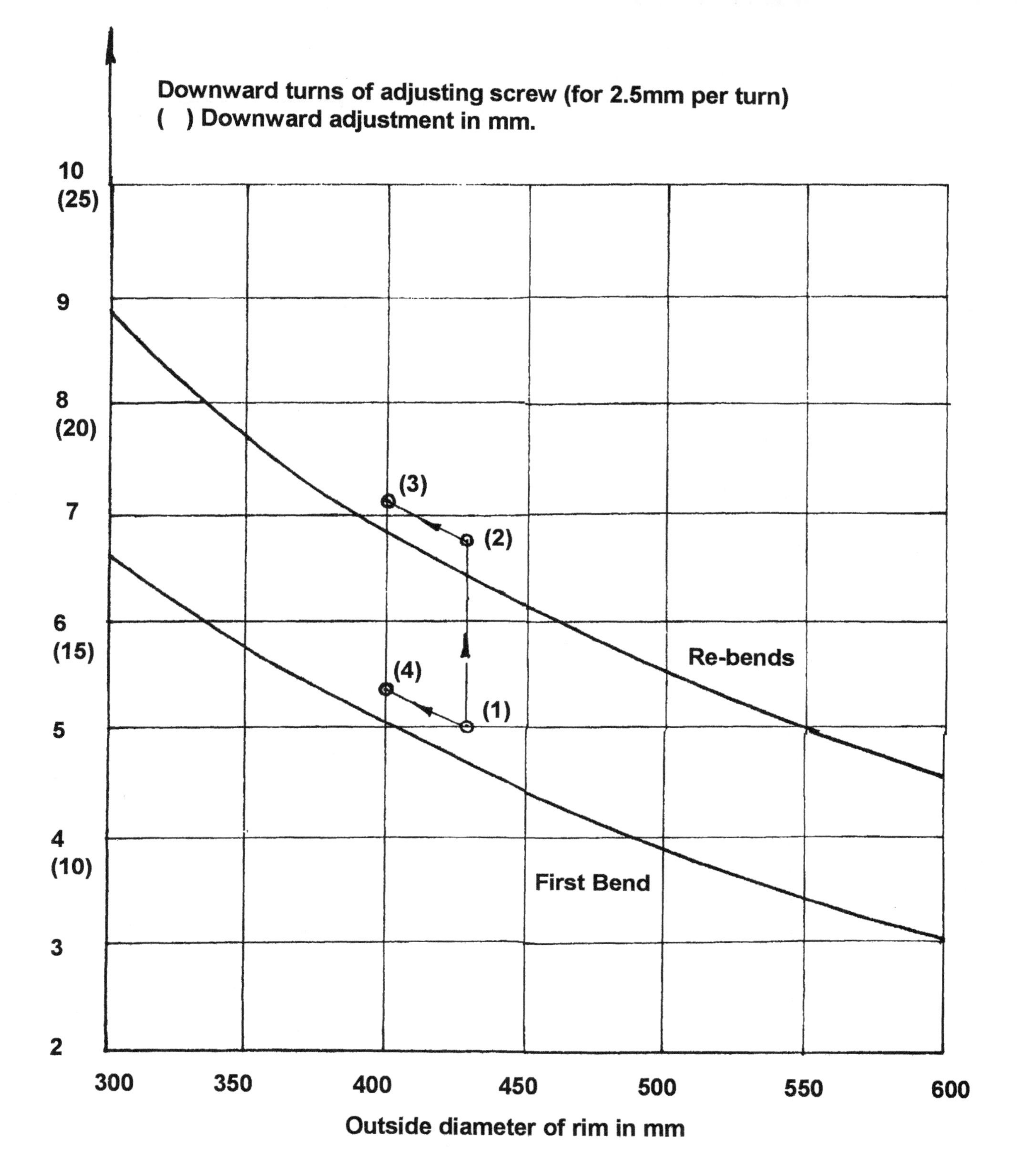


**First Bend Line** - use when rim is bent from straight.
**Re-bend Line** - use for re-bends when rim has already been bent.

Figure 2.2: CHART A - ADJUSTMENT OF RIM BENDER WHEN LOWER ROLLERS ARE IN INNER POSITION

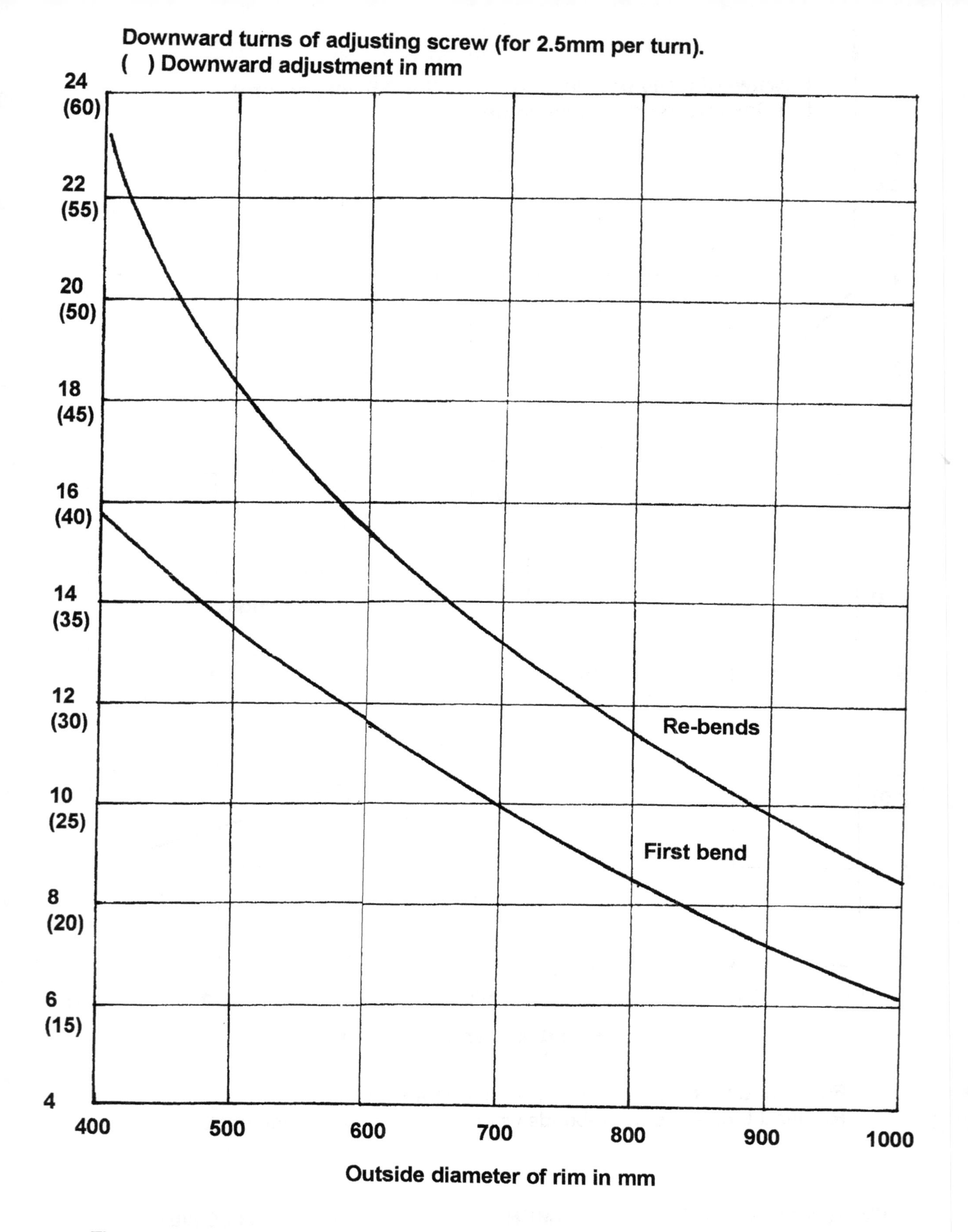


Figure 2.2: CHART B - ADJUSTMENT OF RIM BENDER WHEN LOWER ROLLERS ARE IN OUTER POSITION

# 3. ASSEMBLY JIG

## 3.1 INTRODUCTION

The use of an assembly jig in the production of the wheels is very important:

i) to hold the parts in the correct positions for welding;

ii) to ensure consistent quality of construction in terms of maintaining the correct size and shape of the wheels;

iii) to minimise distortion of the wheel arising from the welding process.

The most important aspect of the jig design is to ensure that the wheel rim is correctly located with respect to the hub so that the wheel runs true with minimum side to side wobble.

The rim needs to be clamped firmly in position so that it is centralised with the hub and square to the hub (i.e. plane of the rim at right angles to the axis of the hub).

Details of the assembly jig are shown in Figure 3.1. It comprises a **base frame** with 8 arms on which are bolted **rim clamps** to hold the rim in the correct position relative to the **centre post** which is fixed at the centre of the base. Each arm has sets of holes so that the rim clamps can be bolted in position for standard size rims from 12" to 18" sizes and also 26" and 28" bicycle tyre sizes.

Each rim clamp has 2 locating holes and 2 slotted holes. A locating hole is used to locate the rim clamp for the standard rim sizes whilst the slotted holes allow some adjustment for other sizes of rims or to cater for variations in tyre size. When the slotted holes are used, a **gauge** should be used to ensure that all the **rim guide posts** are the same distance from the **centre post**. Details are given of the construction of a suitable gauge.

The inside of the rim is clamped against the **guide posts** by the **screw clamps** and the wheel spokes should be fitted as closely as possible to the guide posts so that distortion caused by welding the spokes to the rim is kept to a minimum. The **arms** of the assembly jig are positioned to cater for **4** and **6** spoke wheels - **8** and **12** spoke wheels can be accommodated by welding in half the spokes and then rotating the wheel to allow the fitting of the other half.

It is very important that the hub is centrally mounted on the centre post and it is recommended that machined bushes or collars are used to locate the hub correctly on the post. These collars should also support the hub at the correct height relative to the rim. The collars are the only components which need to be machined on a lathe - if this is not available it is possible to fabricate locating collars for the hub but it would be better to get sets machined up by a workshop which does have a lathe.

The rim should sit evenly on the bases of all the rim clamps. Normally this will occur naturally but if problems arise, 4 posts can be welded to the arms of the base and 2 clamping bars used to clamp the rim down in position.

**Figure 3.1: WHEEL ASSEMBLY JIG**

## 3.2: CONSTRUCTION OF ASSEMBLY JIG

This section gives step-by-step instructions for construction of the assembly jig.

The assembly jig sets the standard of accuracy and quality of construction of the wheels produced with the equipment and therefore it is very important that the jig is accurately made. The instructions emphasise the accuracy of measurement, cutting and setting up parts. Since the assembly jig has 7 clamps to hold the rim in the correct position relative to the hub, it is important that a template (pattern) is used to ensure that the clamping holes are drilled in exactly the same position on each arm and that a jig is used to make sure the clamps are identical.

**Materials:**

The main materials needed to construct the assembly jig are as follows:

50 x 50 x 6 angle section - 4.5m

50 x 6 flat bar - 2.5m

100 x 6 flat bar - 0.1m

16 or 20 diameter round bar - 0.7m

M16 threaded rod (1.0m) and 7 M16 nuts plus 5/8" medium wall pipe (0.5m)
**or**
M20 threaded rod (1.0m) and 7 M20 nuts plus 3/4" medium wall pipe (0.5m)

M8 or M10 bolts x 20 long and nuts - 14 of each.

**Notes:** 50 x 50 x 6 angle section and 50 x 6 flat bar are commonly available sections and these should be used if possible.

20 diameter round bar is the preferred size for the centre-post and for the rim guide-posts. 16, 22 or 25 diameter bar could be used as alternatives.

A combination of threaded rod and tube is needed in which the rod fits closely inside the tube. The two standard sizes which are suitable are given. In the case of the 5/8" tube it may be necessary to file the seam inside the tube for the M16 rod to slide through.

## 1. CONSTRUCTION OF THE JIG BASE

***NOTE:** the accuracy of construction of the jig is **very** important as this will determine the quality of the wheels produced with the jig.

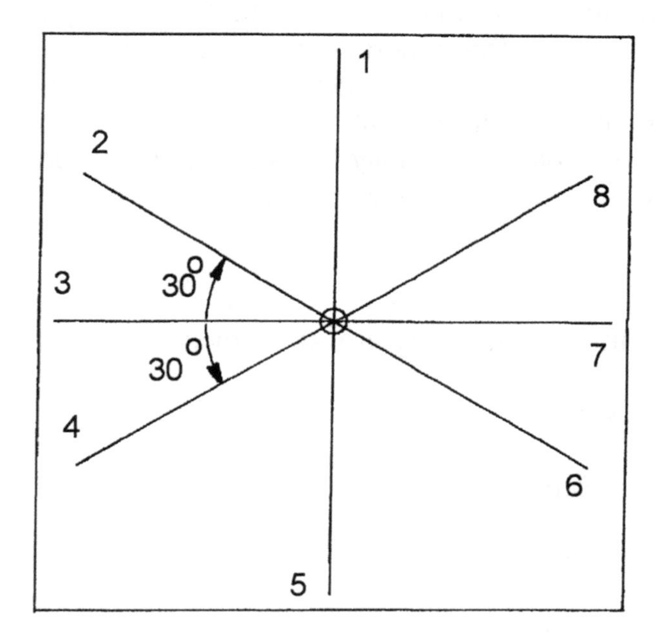


**Step 1: It is recommended that the base is laid out on a flat board - for example 5/8", 3/4" or 1" plywood - about 1m square**

**Note:** that the base is assembled **upside down** on the board.

**1.1:** Locate the centre of the board and carefully mark out and draw radial lines for the centre-lines of the arms of the jig.

**1.2:** Drill a **10mm** diameter hole at the centre of the arms

### Step 2: Cut the arms of the jig

**2.1:** Cut a length of **50 x 50 x 6 angle, 850mm long.** This will be arms **1** and **5** of the jig.

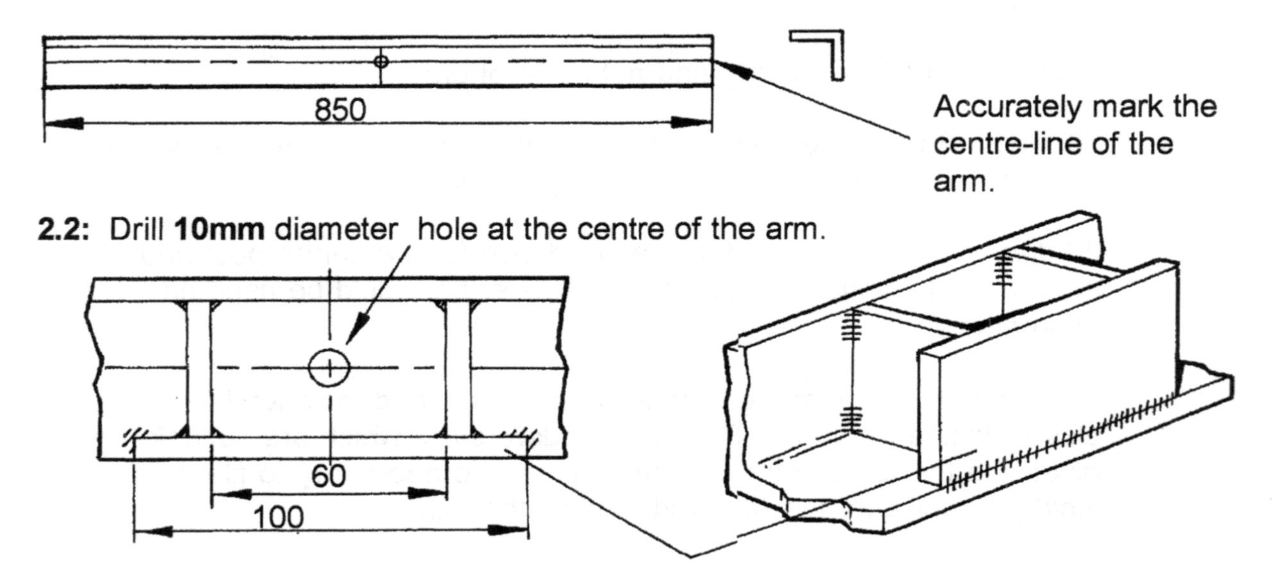


**2.2:** Drill **10mm** diameter hole at the centre of the arm.

**2.3:** Reinforce the centre of the arm with pieces of **6mm** thick flat bar as shown.

**2.4:** Cut the other arms from **50 x 50 x 6** angle

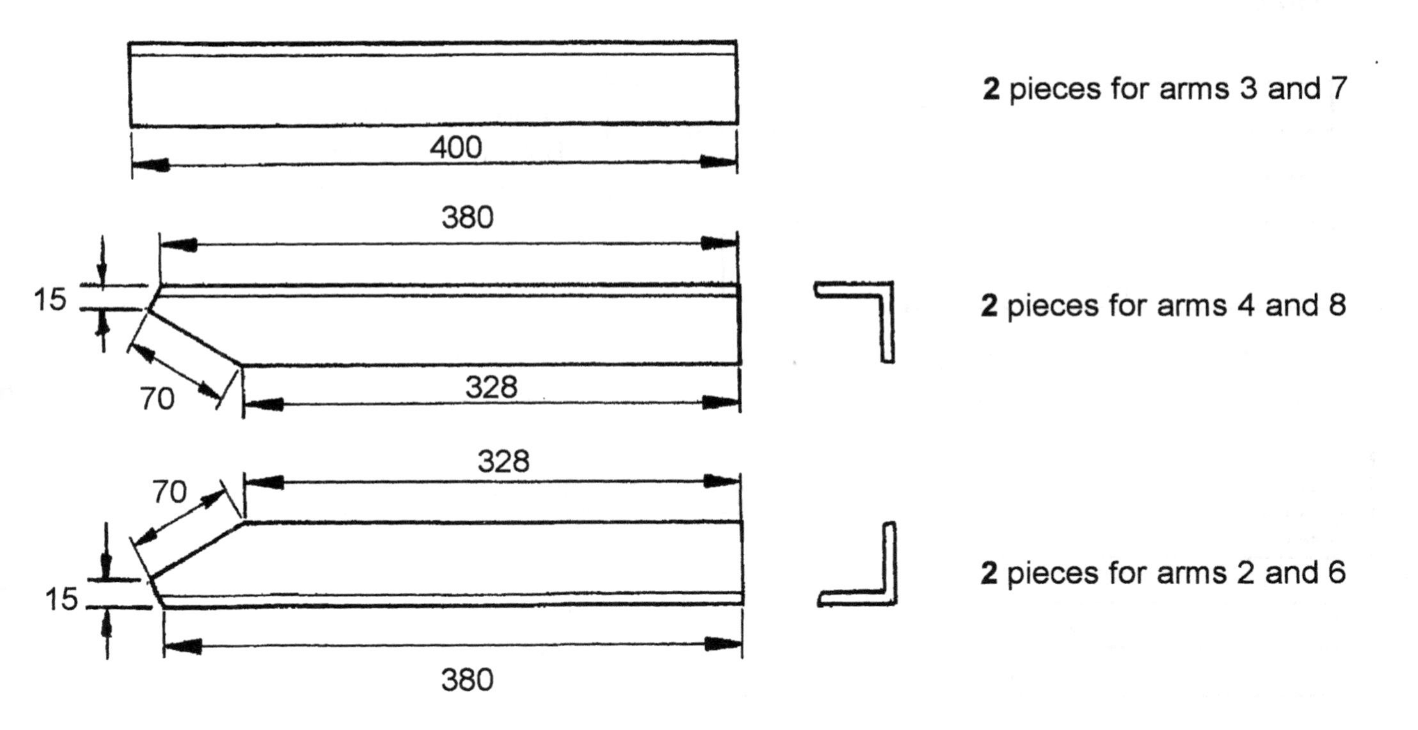


**Step 3: make a drilling template for the arms**

**3.1:** Cut a piece of **50 x 6** flat bar, **450** long

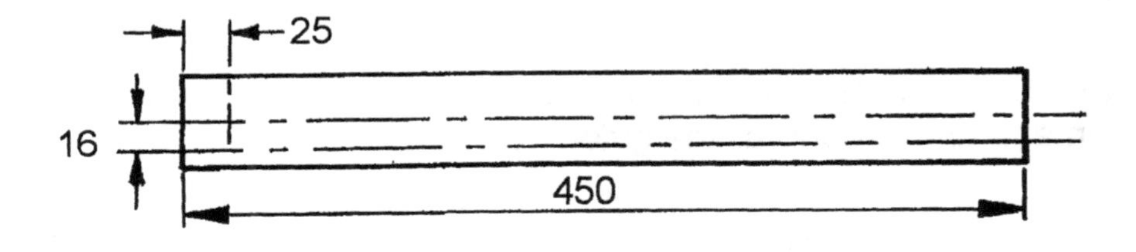


**3.2:** Carefully measure and mark the centre-line along the length of the bar and a second line 1**6mm** from the centre-line

**3.3:** Mark a point **25mm** from one end. This point will be used to line up with the centre of the base frame

**3.4:** Carefully measure and mark the positions of the holes as shown on the following drawing.

**3.5:** Drill **10 diameter** for centre location and **8 diameter** for **M8** bolts or **10 diameter** for **M10** bolts at all other positions.

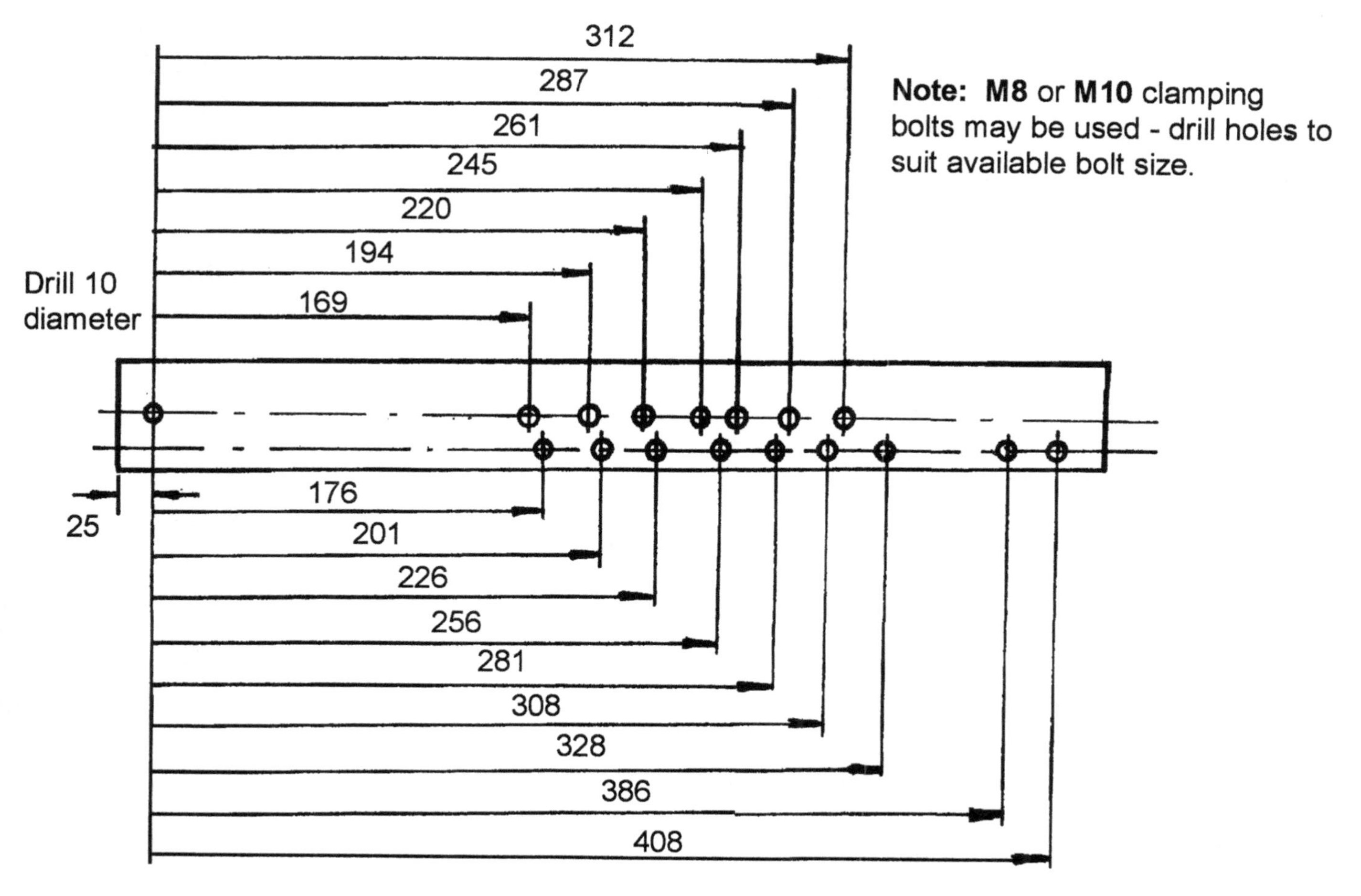


**Step 4: Use template to drill holes in arms**

**4.1:** Bolt the template to arm **1-5** at the centre. Line up the template with the edges of the arm. Clamp it in position and drill through all the guide holes.

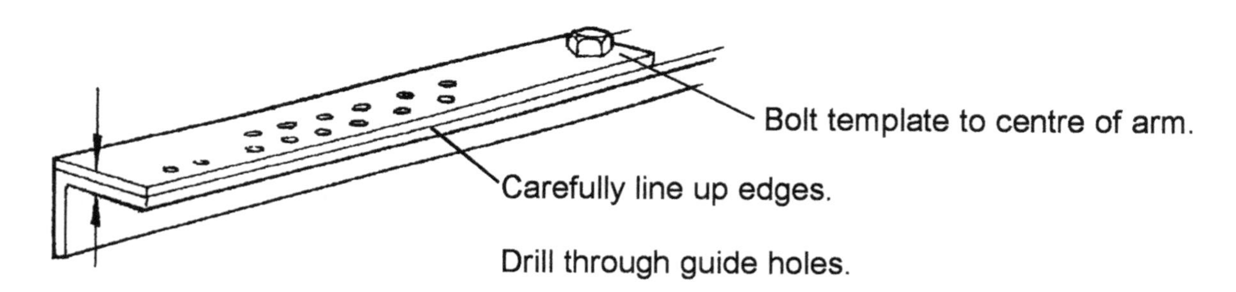


**Note:** the **offset** row of holes must always be on the **outer** edge of the angle - not the corner edge.

**4.2:** Remove the bolt, turn the template **over** and bolt it at the other end of the arm. Line it up, clamp it in position and drill through the guide holes.

**4.3:** Drill the other arms of the jig.

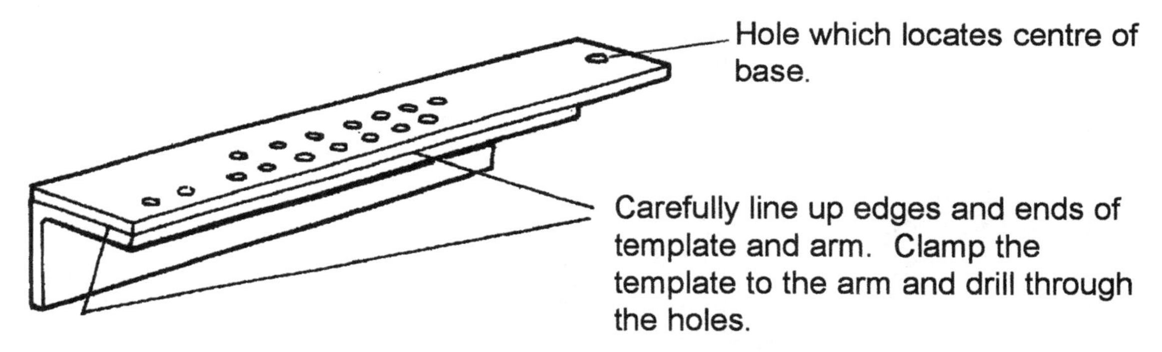


**Note:** The **off-set** row of holes will always be on the **outer** edge of the angle not the corner edge.

**Step 5: Drill locating holes for the arms in the layout board**

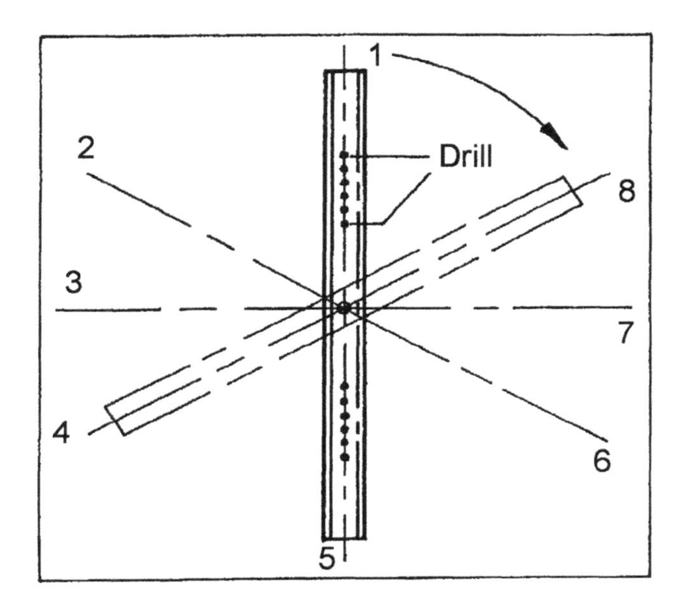


**5.1:** Bolt arm **1-5** to the board through the central hole.

**5.2:** Line up the centre-line of the arm with line **1-5** on the board.

**5.3:** Clamp the arm in position and drill these two holes through the board at each end of the arm.

**5.4:** Rotate the arm so it lines up with line **4-8** on the board and drill through the same pair of holes at each end of the arm.

**5.5:** Repeat the procedure for arms **3-7** and 2-6.

### Step 6: Assemble the base frame on the layout board

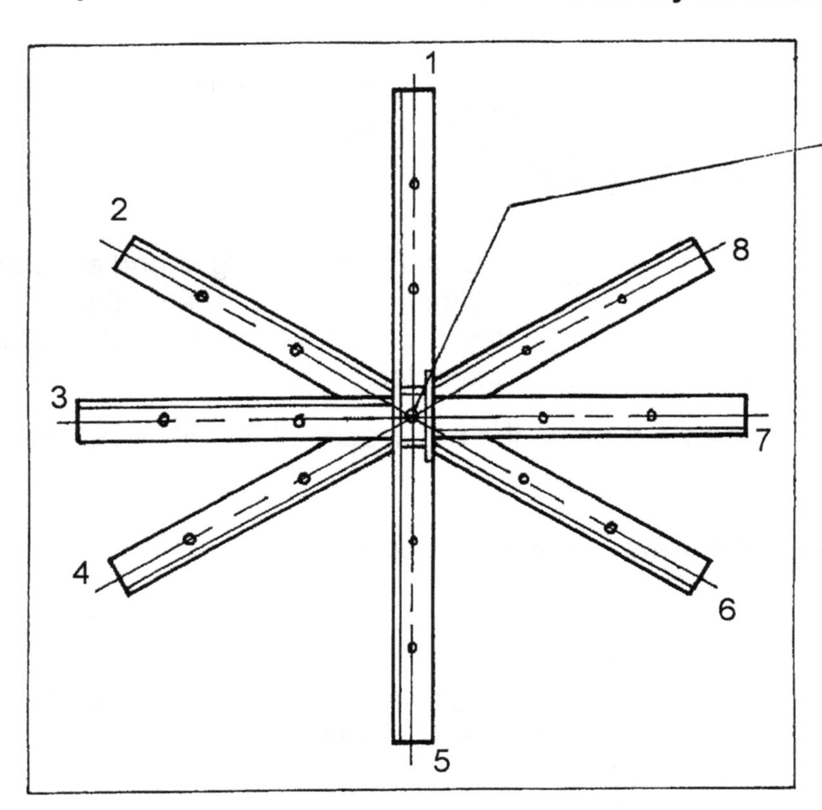


**6.1:** Bolt arm **1-5** at the centre of the board and one of the outer holes.

**6.2:** Bolt each of the other arms to the board at the locating holes.

The arms should fit together with gaps of 1-2mm at the joints to allow for welding.

**6.3:** Join the arms together with strong tack-welds.

**Note:** the frame is assembled upside-down.

**6.4:** Remove the frame from the board and complete the welding. Try to balance the welds on top and bottom of the frame as much as possible to minimise bowing of the frame.

**6.5:** When welding is complete, grind the welds on top of the frame flat and check the flatness of the top of the frame with a straight-edge.

**6.6:** If there is any bowing of the frame use a press, jack or hammer to straighten it out.

**Note:** the accuracy of the jig and therefore the quality of wheels produced depend very much on the flatness of the base frame and the correct positioning of the bolt holes. Therefore it is very important to get it right.

**Step 7: Fit the centre-post at the centre of the base-frame**

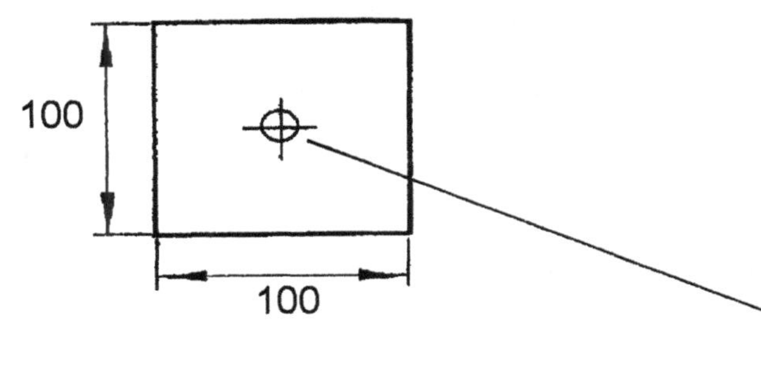


**7.1:** Cut a flat plate from **100 x 6** flat bar (or weld together **2** pieces of **50 x 6** flat bar - grind the welds flush on both sides and flatten the plate).

**7.2:** Mark the centre of the plate and drill a **10 diameter** hole.

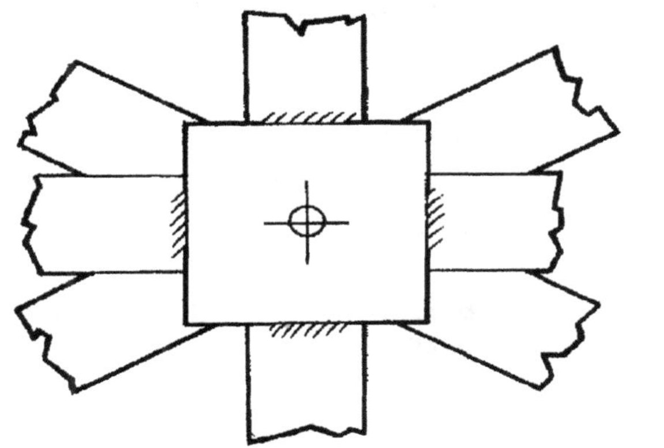

**7.3:** Use an **M10** bolt to clamp the plate centrally on top of the base-frame.

**7.4:** **Check** that the plate is flat with the top of the base-frame and weld the plate in position.

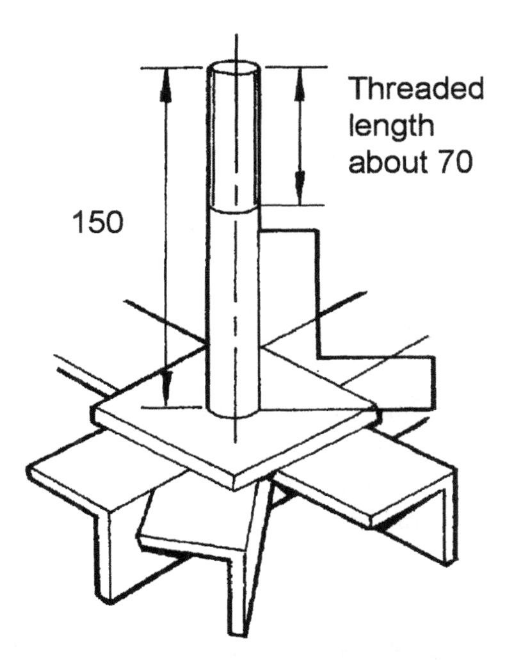


**7.5:** Cut a piece of bar **170** long for the centre post. This may be **12, 16 or 20 diameter** depending on what size of bar and/or drill size is available - 16 or 20 diameter is preferred. The top **70mm** should be threaded.

**7.6:** Drill out the centre hole to the centre-post diameter and fit the centre-post as shown. **Check** that the centre-post is square to the base **in all directions.**

**7.7:** Turn the base on its side and weld the post in position **underneath** the base.

**Note:** It is very important that the post is square to the base. After welding recheck it and if it is not square use a piece of pipe to bend it to the correct position.

**Step 8: Make attachments to clamp rim down on base**

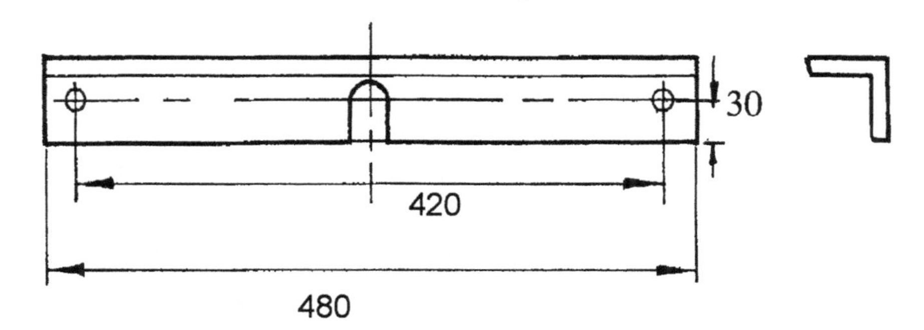


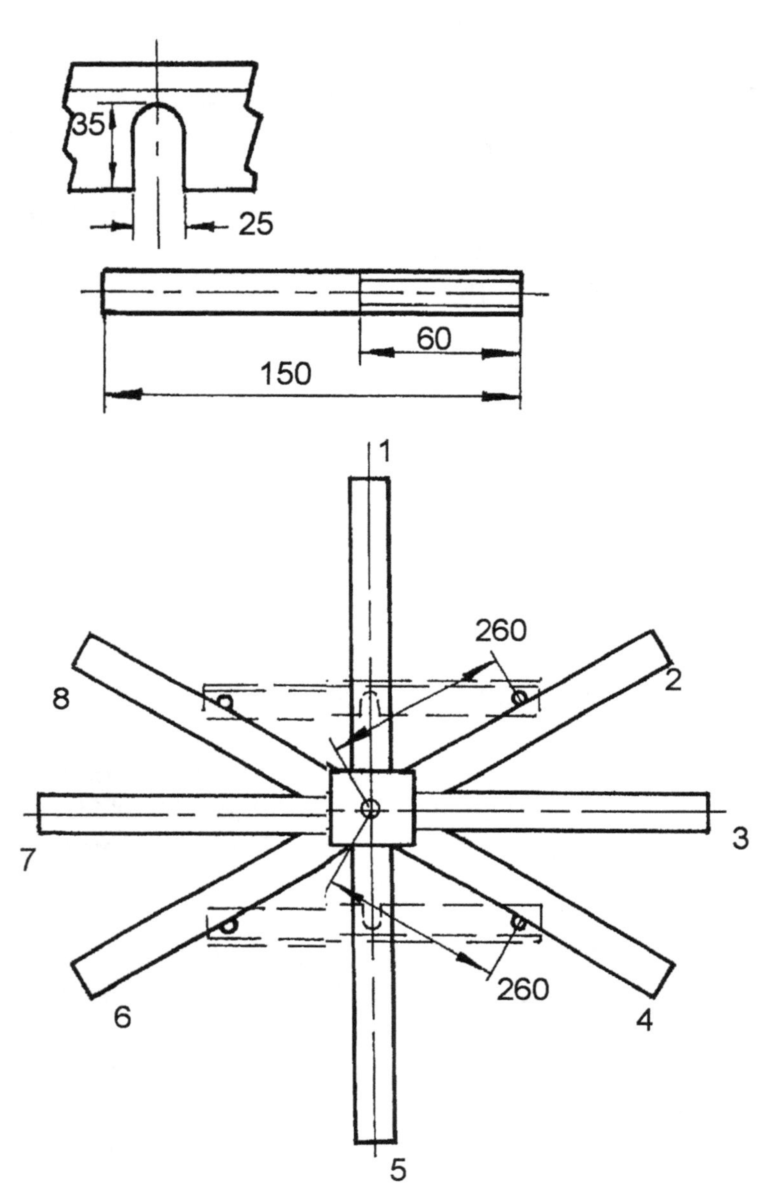


**8.1:** Cut **2** pieces of **50 x 50 x 6** angle **480** long.

**8.2:** Drill **2** holes for **10 or 12** diameter posts, **210** either side of centre.

**8.3:** Cut a central slot, **25** wide **x 35** deep.

**8.4:** Cut **4** posts from **10 or 12** diameter bar, threaded if available. At least top **60mm** should be threaded

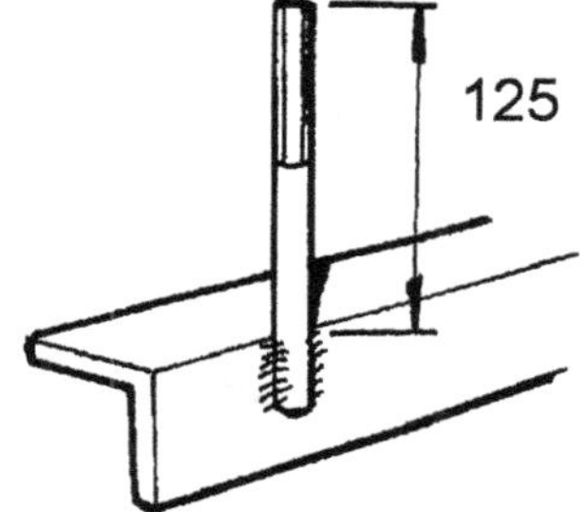


**8.5:** Locate 2 posts on arms **2** and **4** as shown. Check they are square to the top surface of the arm and weld them in position.

**8.6:** Fit the clamping bars over the posts and locate the other two posts on arms **6** and **8.** Check the posts are the correct height and square to the base and weld them in position

## 2. CONSTRUCTION OF 7 RIM CLAMPS

These clamps bolt on to the arms of the base and are used to clamp the rim in position. The accuracy of construction is very important. A jig should be made to ensure that all 7 clamps are made identical.

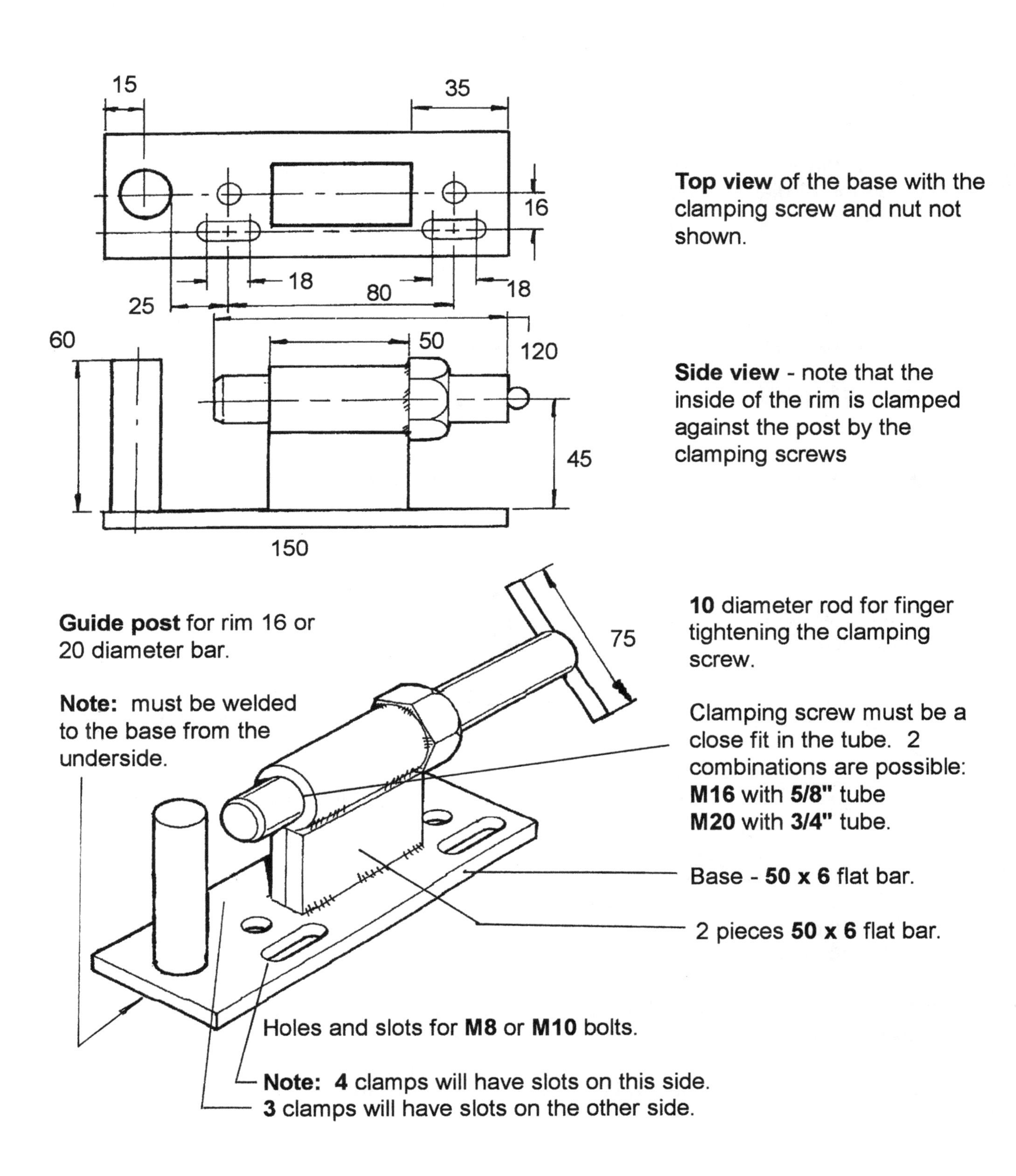

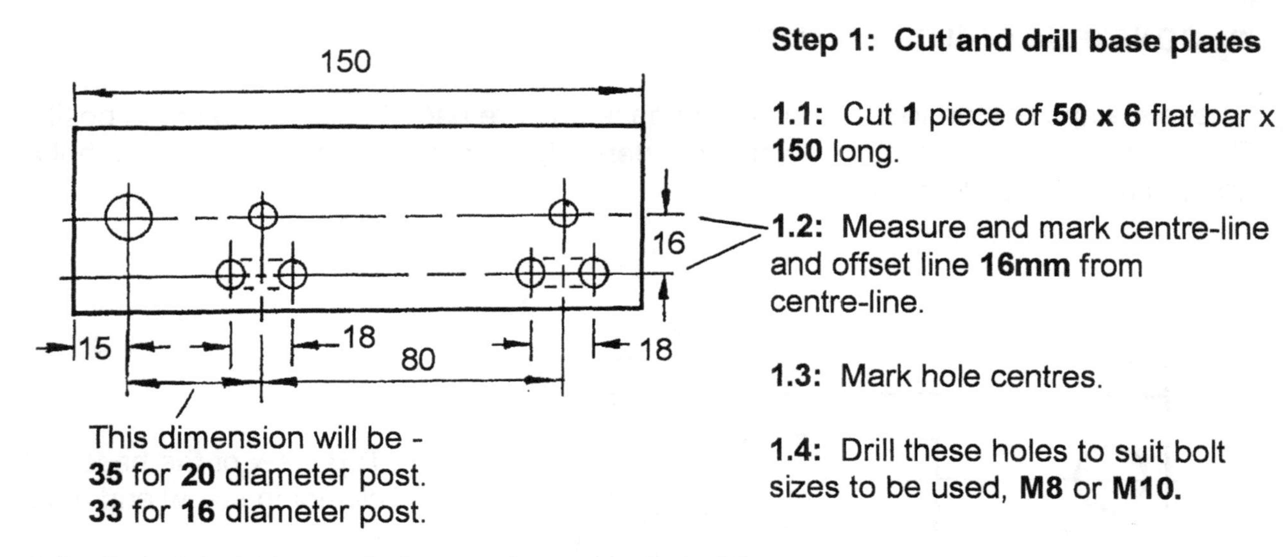


**Step 1: Cut and drill base plates**

**1.1:** Cut **1** piece of **50 x 6** flat bar x **150** long.

**1.2:** Measure and mark centre-line and offset line **16mm** from centre-line.

**1.3:** Mark hole centres.

**1.4:** Drill these holes to suit bolt sizes to be used, **M8** or **M10.**

**1.5:** Cut slots between holes as shown by dotted lines.

**1.6:** Check that space between holes and slots is same as between two rows of holes on base arms. 1 hole and both slots should line up with holes in arm.

**1.7:** Drill hole for post either to size or to allow welding of post from underneath base.

**1.8:** Cut **6** more pieces of **50 x 6** flat bar and use above piece as a template to drill the 6 pieces.

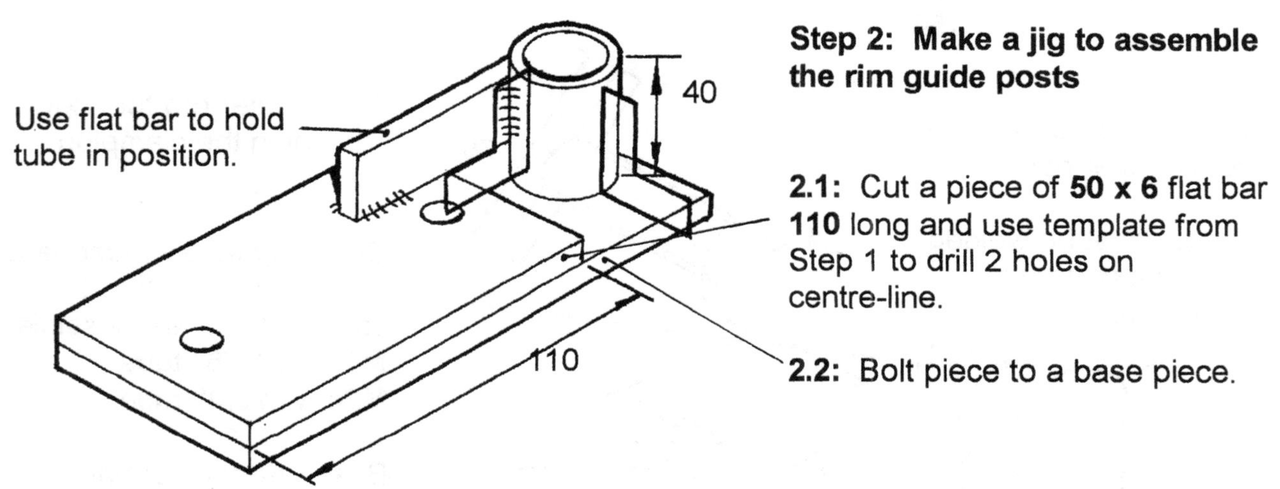


**Step 2: Make a jig to assemble the rim guide posts**

**2.1:** Cut a piece of **50 x 6** flat bar **110** long and use template from Step 1 to drill 2 holes on centre-line.

**2.2:** Bolt piece to a base piece.

**2.3:** Cut a **40mm** length of tube which will fit closely over a post:
**5/8"** for **16** diameter post
**3/4"** for **20** diameter post

**2.4:** Clamp the tube to the base so that it fits centrally over the post hole. Make sure the tube is square to the base.

**2.5:** Cut a piece of flat bar and weld as shown to hold the tube in the correct position.

**Step 3: Use jig to assemble the posts on the base**

**Completed Jig**

**3.1:** Bolt the jig on top of the base piece.

**3.2:** Cut **7** pieces of round bar for the posts **16** diameter or **20** diameter x **70** long.

**Note: 20** diameter is preferred.

**3.3:** Fit the post in the locating tube and weld it to the base from the **underside.**

**3.4:** Repeat steps 3.1 to 3.3 for the other 6 bases.

**Note:** **4** bases must have slots on **this side.**
**3** bases will have slots on the **other side.**

**Step 4: Make a jig to assemble the screw clamp**

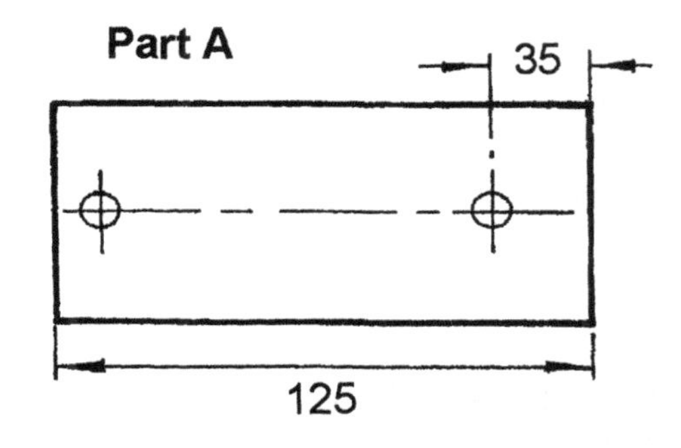


**4.1:** Cut **1** piece of **50 x 6** flat bar x **125** long and use the template from Step 1 to drill 2 holes on the centre-line.

**4.3:** Cut length of threaded rod, **M16** or **M20.** Position it on flat bar, check that it is square and weld it to bar.

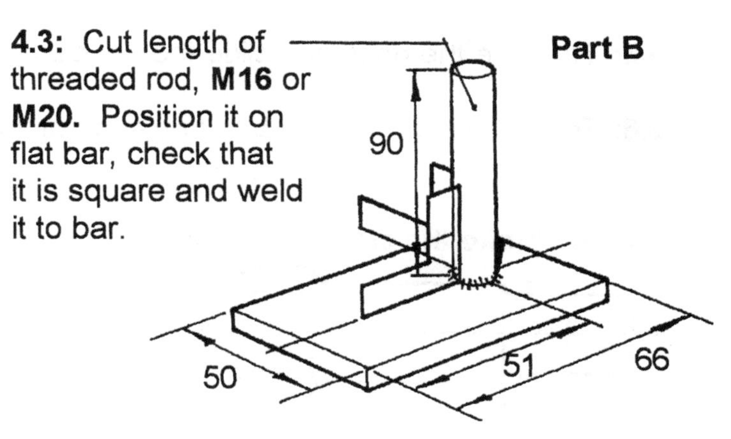


**4.2:** Cut piece of **50 x 6** flat bar x **66** long.

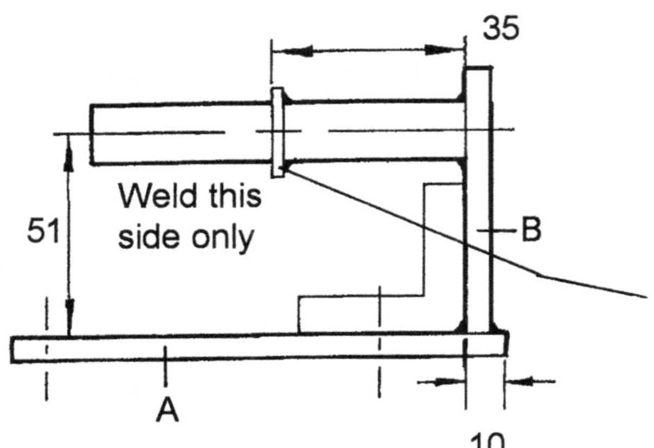


**4.4:** Position part **B** on part **A** as shown. Check that B is square with A and that the centre of the threaded rod is at the correct **51mm** above A. Weld B to A.

**4.5:** Weld a washer squarely on the threaded rod as shown.

**Step 5: Use the jig to assemble the tube for the screw clamp**

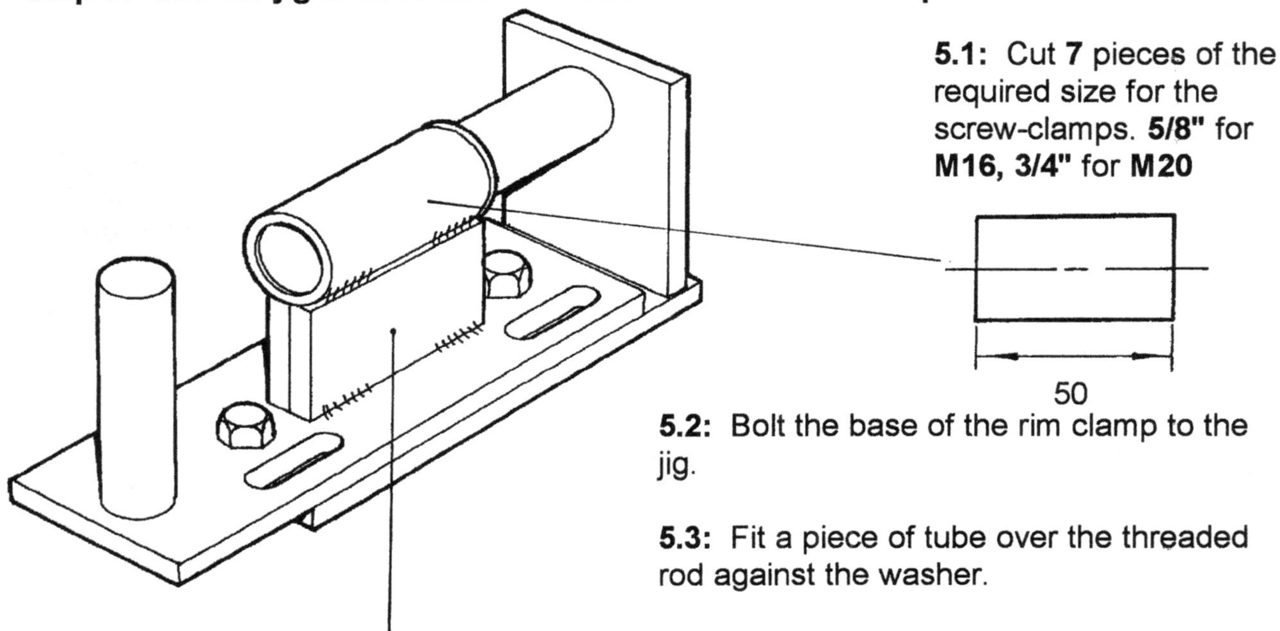


**5.1:** Cut **7** pieces of the required size for the screw-clamps. **5/8"** for **M16, 3/4"** for **M20**

**5.2:** Bolt the base of the rim clamp to the jig.

**5.3:** Fit a piece of tube over the threaded rod against the washer.

**5.4:** Measure and cut **2** pieces from **50 x 6** flat bar. Fit these under the tube either side of the centre-line and weld them to the base and to the tube.

**5.5:** Remove the bolts and slide the rim clamp off the jig.

**5.6:** Repeat Steps 5.2 to 5.5 to make the other 6 clamps.

**Step 6: Make the clamping screws**

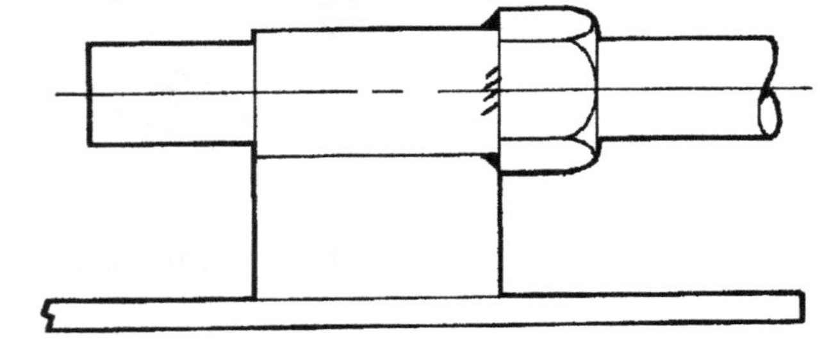

**6.1:** Cut **7** lengths of threaded rod x **120** long.

**6.2:** Screw the nut onto a piece of rod and slide the rod through the tube. Weld the nut to the tube.

**6.3:** Repeat this step for the other 6 clamps.

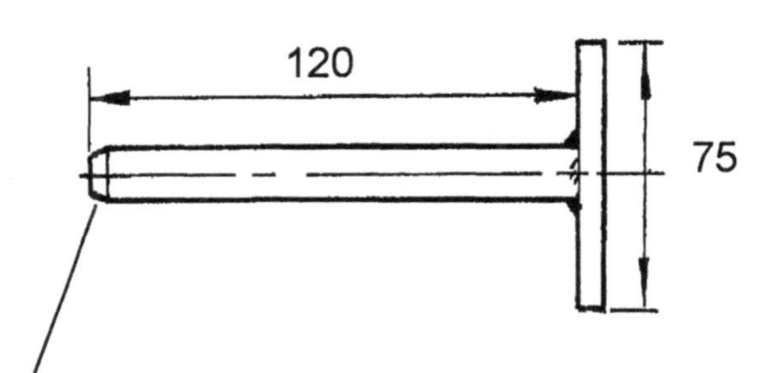


**6.4:** Weld a piece of **10 diameter** rod across the end of each piece of threaded rod.

**6.5:** Grind or file an **even** chamfer onto the end of each screw. This is so it will seat evenly inside an angle section.

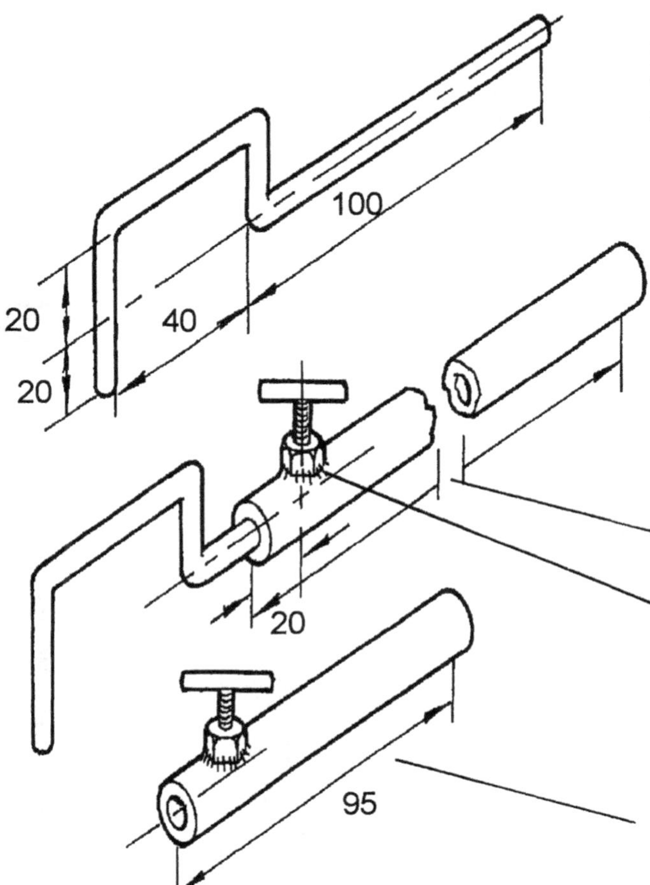


**Step 7: Make gauges for setting the positions of the rim clamps.** (These are for use when the clamps are not located by fixed holes on the base arms).

A combination of rod and tube is needed so that the rod will slide neatly in the tube. The rod may be 6 diameter or larger.

**7.1:** Bend the rod to the shape shown.

**7.2:** Cut the tube to a length of **205mm**.

**7.3:** Drill a **7** diameter hole and weld an **M6** nut over the hole.

**7.4:** Make a finger-tightening screw with an **M6** bolt.

**7.5:** Make a tube up for smaller size wheels.

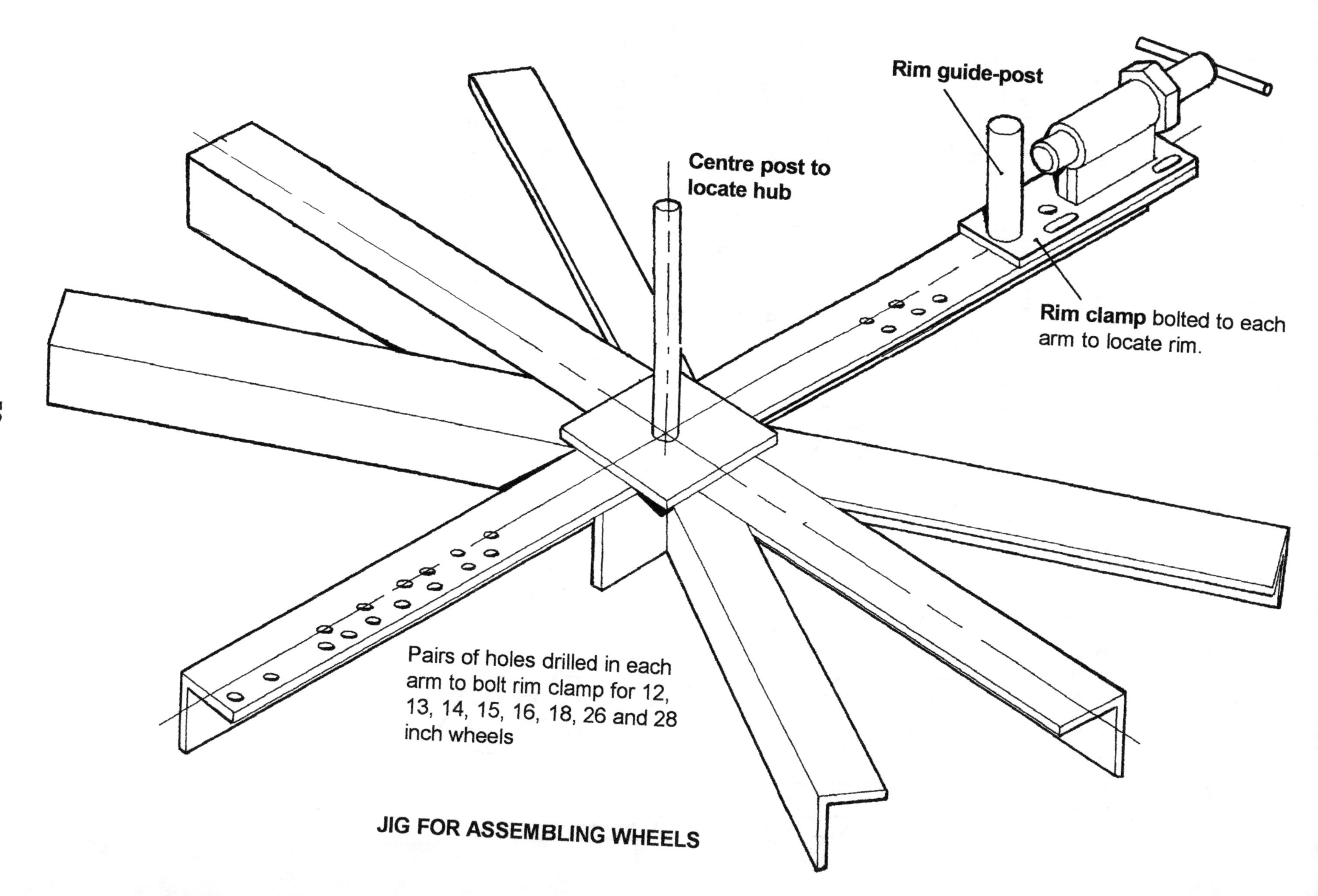


**JIG FOR ASSEMBLING WHEELS**

## RIM CLAMPING POSITIONS

The following tables list the range of rim diameters which can be set up with the rim clamping positions provided on the assembly jig.

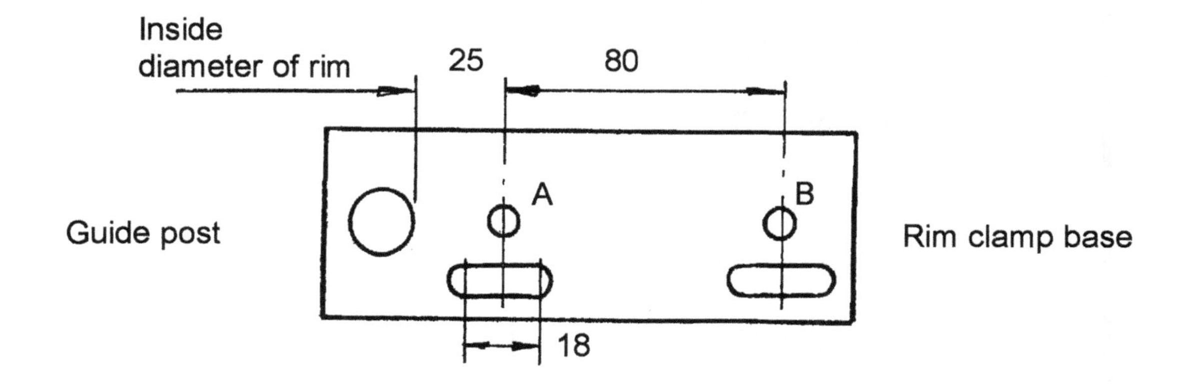


### 1. Fixed positions using holes A or B

| Wheel size | Hole used on clamp | Radius of hole used on arm of base (mm) | Inside diameter of rim (mm) | Outside diameter of rim for 6mm thick rim (mm) |
|---|---|---|---|---|
| 12" (304.8mm) | A | 169 | 288 | 300 |
| 13" (330.2) | B | 261 | 312 | 324 |
| 14" (355.6) | A | 194 | 338 | 350 |
| 15" (381) | B | 287 | 364 | 376 |
| 16" (406.4) | A | 220 | 390 | 402 |
| 17" (431.8) | B | 312 | 414 | 426 |
| 18" (457.2) | A | 245 | 440 | 452 |

## 2. Using Slotted Holes

| Radii of holes used on arm of base (mm) | Range of inside diameter of rim (mm) |
|---|---|
| 176 and 256 | 284 to 320 |
| 201 and 281 | 334 to 370 |
| 226 and 308 | 388 to 424 |
| 256 and 328 | 444 to 464 |
| 308 and 386 | 544 to 580 |
| 328 and 408 | 588 to 624 |

**Some of rim sizes that can be assembled using the rim clamps bolted through the slotted holes**

| Wheel size | Inside diameter of rim |
|---|---|
| **1. Bicycle-type wheels with 25 x 25 angle rim** | |
| 20" | 414mm |
| 26" | 565mm |
| 27" | 591mm |
| 28" | 616mm |
| **2. Motorcycle type wheels with 40 x 40 angle rim** | |
| 17" | 400mm |
| 18" | 424mm |

**Note:** that once a correct rim size has been set using the slotted holes, additional holes can be drilled in the base arms through holes A or B to fix the rim clamps in position for future use.

### 3.3: USE OF THE ASSEMBLY JIG

**Step 1: Bolt the rim clamps to the arms of the jig in the correct positions for the size of rim required**

**1.1** For **4** spoke wheels use arms **1,2,3,5,7.** The rim joint should be mid-way between arms **2 and 3.**

For **6** spoke wheels use arms **1,2,3,4,5,6,8.** The rim joint will be mid-way between arms **2 and 3.**

**1.2** Holes are provided to locate the rim clamps in the correct positions for car and motorcycle tyre sizes of the following range: 12", 13", 14", 15", 16", 17", 18". In each case one locating hole is used and one slotted hole on the rim clamp base as shown below:

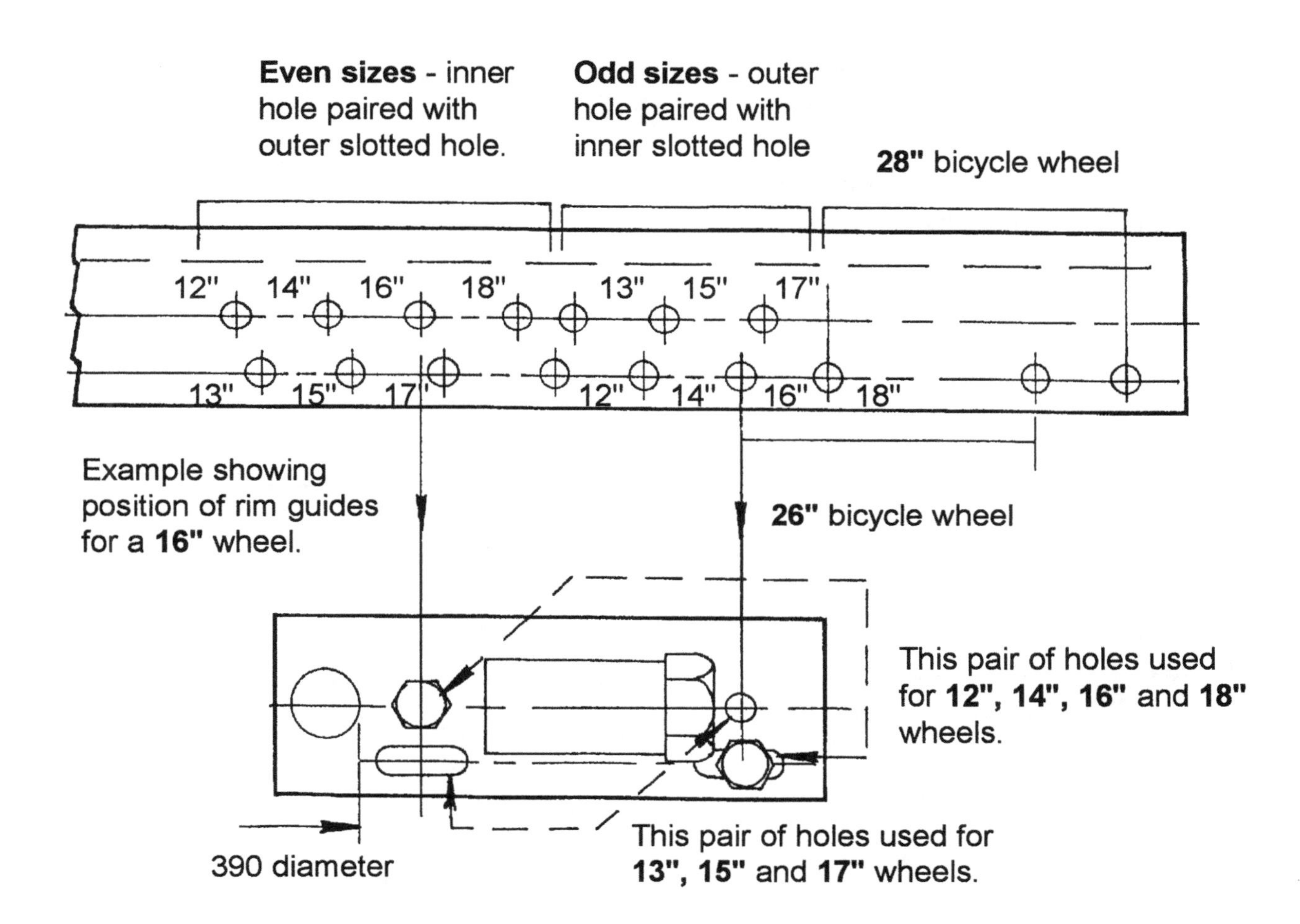

Note that the locating holes are for rim thicknesses of 6mm. For other thicknesses of rim, pieces of packing may be used between the post and the rim, or the pair of slotted holes on the clamp base may be used and the position of the rim guide-post set using the gauge.

The fit of bicycle tyres is quite critical - they must be as **tight as possible** on the rim.

The slotted holes are therefore used to allow some adjustment but when a satisfactory rim size has been obtained holes may be drilled in the arms to exactly locate the rim guides.

**Step 2: Clamp the rim in the jig**

**2.1** When the rim clamps have been correctly positioned and tightly bolted to the arms, the rim is fitted over the guide posts and the screw clamps used to clamp the rim against the guide posts.

**Note:** that the tightening sequence of clamps should start on the arm opposite the joint of the rim and work towards the joint i.e. **7,6,8,5,1,3,2.**

**2.2** Methods for clamping different types and widths of rim are shown in the sketches below.

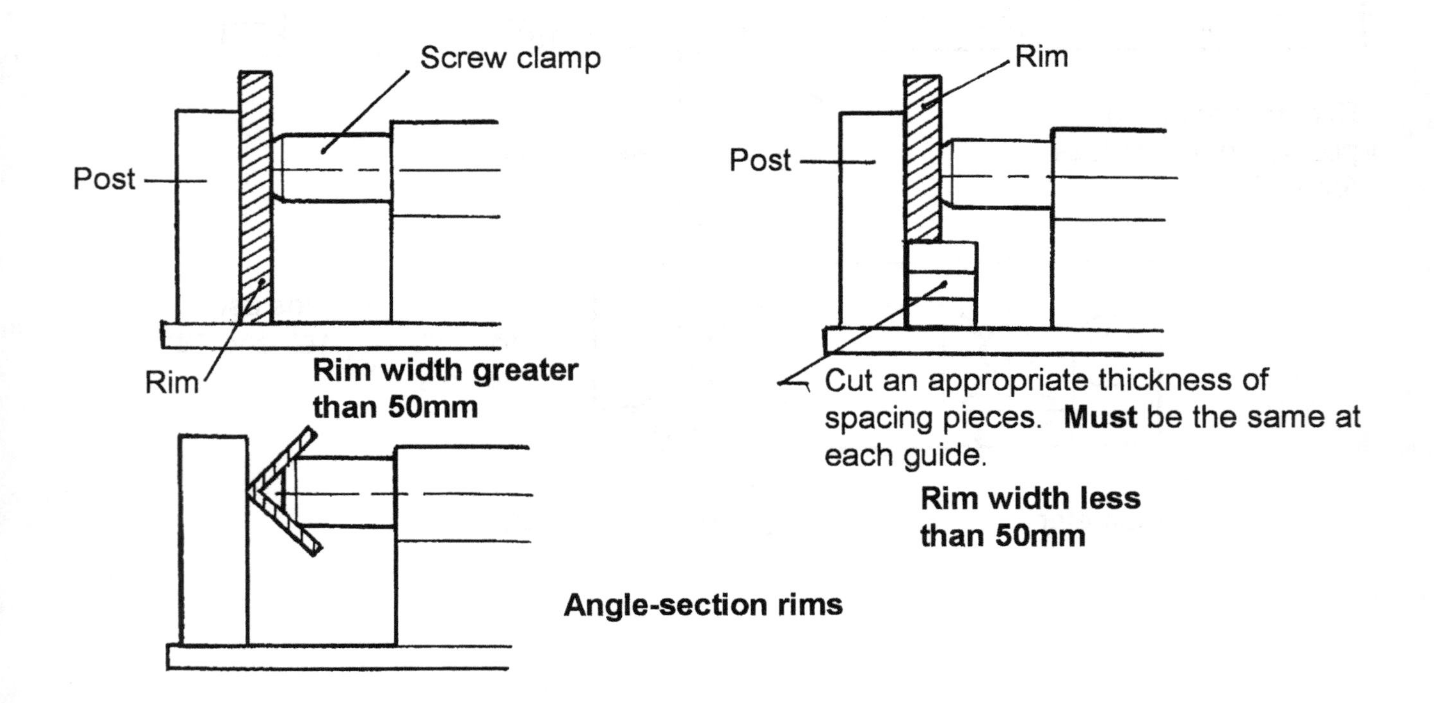

**2.3** If the rim does not sit evenly on all the rim clamps use the two clamping bars to clamp it down in position. However, this should not normally be necessary.

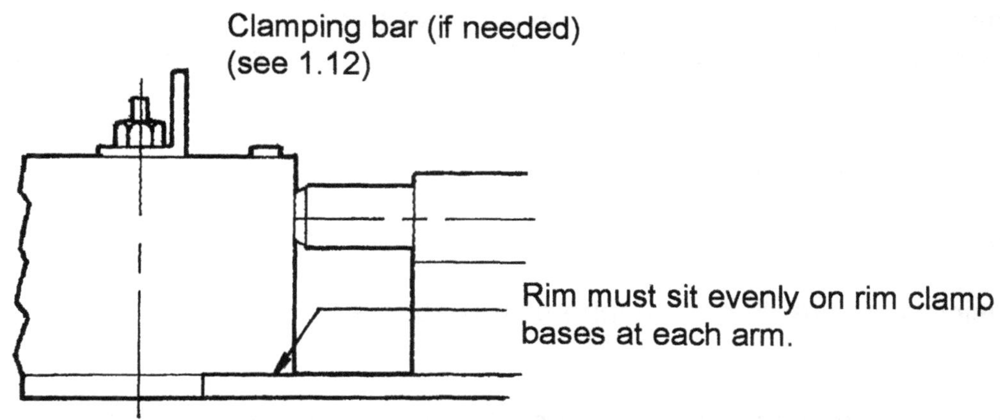


**2.4** When the rim is properly clamped weld the joint between the two ends.

**Step 3: Fit the hub on the centre post.**

**3.1** It is **very important** that the wheel hub is accurately located on the centre post of the jig, and it is recommended that sets of locating collars are machined on a lathe.

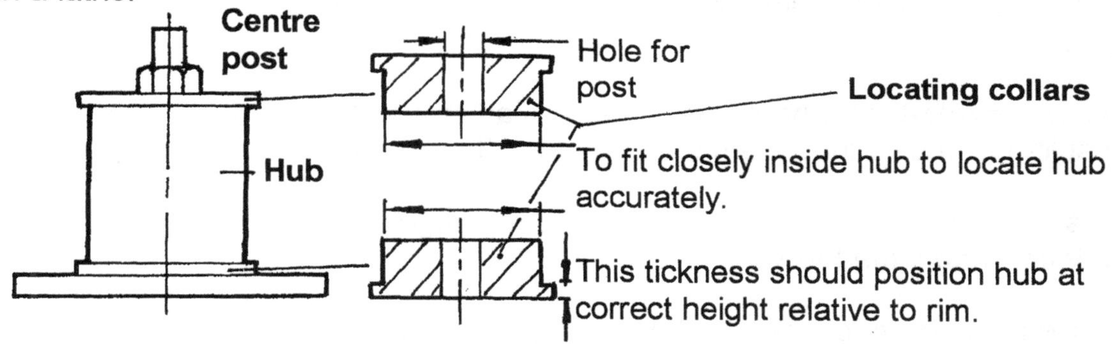


Locating collars may also be fabricated but great care is needed to ensure they are made accurately so that the hub is centrally positioned on the centre post.

**Note:** It is also very important that the ends of the hub are made square otherwise when it is clamped in position it may bend the centre post and pull the hub out of square. This will lead to excessive wobble of the wheel.

For this reason it is a good idea not to clamp down too tightly on the hub.

### 3.3: Wheel wobble

The shape and true-running of the wheel are dependent mainly on the accurate construction of the assembly jig and the correct setting up of the wheel in the jig.

Because wheels are welded some distortion or out of trueness is inevitable but this can be kept well within acceptable limits by correctly setting the wheel up in the jig.

The important factors to be checked are:

- Check, using a straight edge, that the base of the jig is flat
- Check, using a square, that the centre post is square to the base in all directions
- When the rim is clamped in the rig it should be the same height above the base all the way around (providing the width of the rim materials is uniform)
- The ends of the hub must be made square
- Spokes must be cut to the correct length - there should be a gap of 1 or 2mm at each end for welding. If the spokes are too short the extra welding to fill up the gaps will pull the rim out of shape. Spokes should not be forced into position.

Because **bicycle-type** wheels are larger in diameter and have more flexible rims than split-rim wheels they are more likely to suffer from sideways wobble.

If the side-to-side movement of the wheel is too large when it rotates - greater than 6mm - make the following check:

1. Make a square to check that the centre-post is square to the plane of the rim across each set of rim guides.

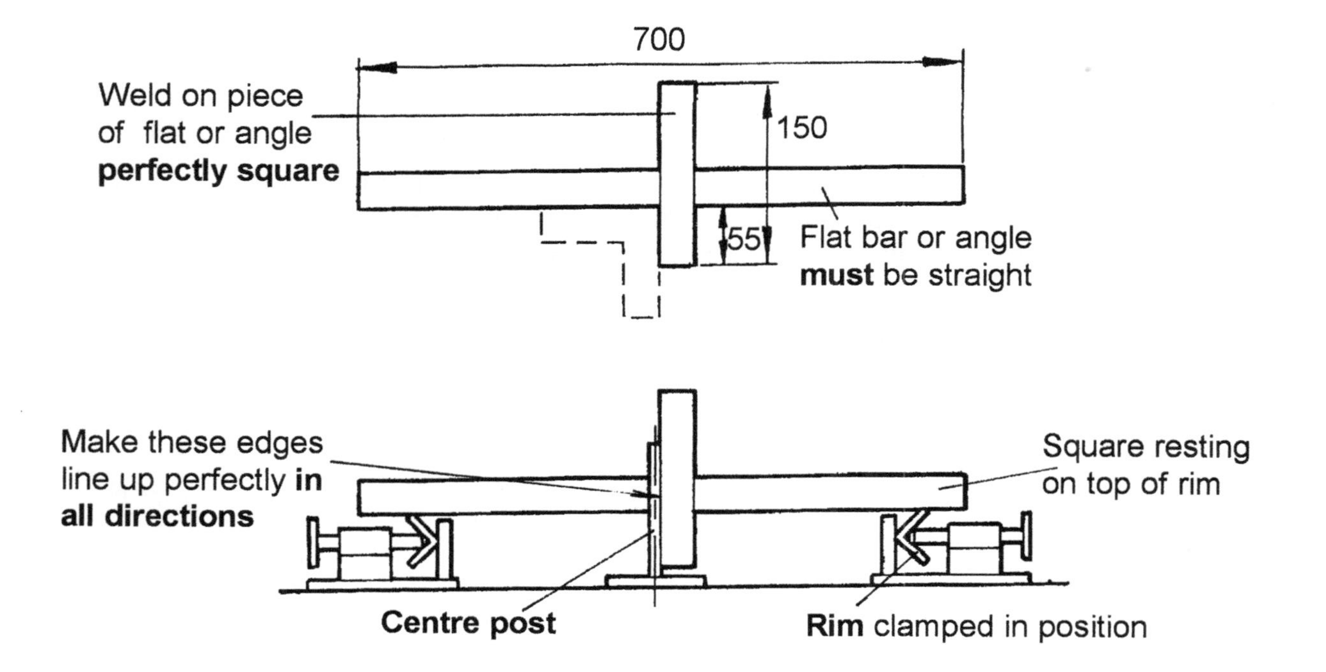

# 4. WHEEL-AXLE ASSEMBLIES

The wheel manufacturing equipment can produce a variety of different wheel designs to suit different applications. The wheel rim has to be designed to take the type of tyre, pneumatic or solid rubber, to be used. It is usually formed from angle section or flat bar and is bent to the required size using the wheel rim bender. The rim is attached to the wheel hub (or axle in live-axle assemblies) usually by spokes, although a plate or disc centre can also be used in some designs. The rim, spokes (or disc) and hub (or axle) are welded together in the assembly jig to ensure that wheels are consistently made to a good level of accuracy and quality.

The wheel either rotates on a fixed axle on bearings fitted in the hub, or may be fixed to an axle which rotates in bearings fixed to the frame of the vehicle (live-axle). The hub, bearings and axle are therefore integral parts of the wheel design and have a very important influence on the efficiency and service life of the wheel assembly. This Chapter therefore describes various axle assemblies which may be used with the wheels produced with the wheel manufacturing equipment. It gives details of the different ways in which the axle may be supported from the vehicle frame, describes various types of bearings which may be used and outlines methods of keeping sand and dust out of the bearings. The general factors which need to be considered in choosing a wheel/axle assembly for a particular application are discussed. Details of specific designs are given in Chapters 5 and 6.

## 4.1: BEARING/AXLE ASSEMBLIES

There are five basic assemblies for mounting the wheel on the vehicle frame:

1. **Fixed Axle supported at both ends** (bicycle-type axle)

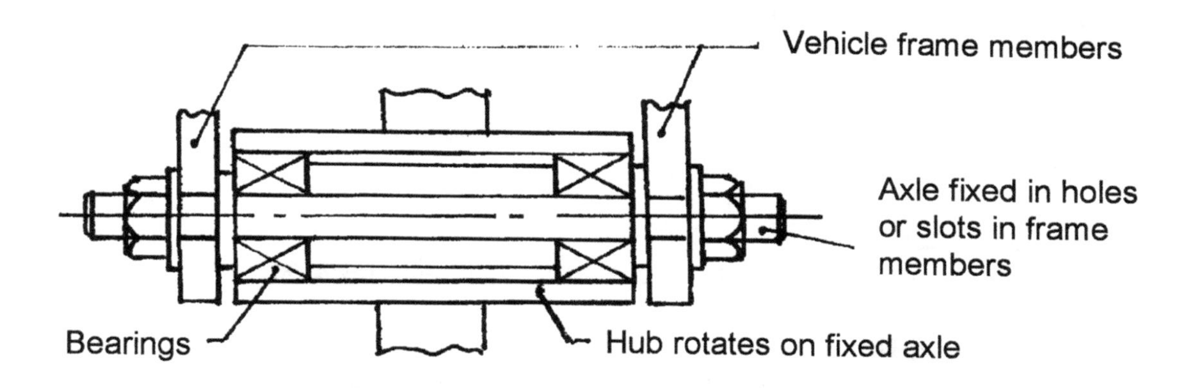

The bearings are mounted in the hub of the wheel and rotate with the wheel on a fixed axle - the axle is supported at both ends.

**Advantages:**
reduces bending of the axle and allows a smaller axle size to be used.

**Disadvantages:**
more difficult to mount and requires more complex design of vehicle frame to support the wheels on both sides.

**Recommended Uses:**
for carts and trailers if a bicycle or motor-cycle wheel axle is used and for wheelbarrows.

**2. Fixed Axle supported at inner end only** (stub axle)

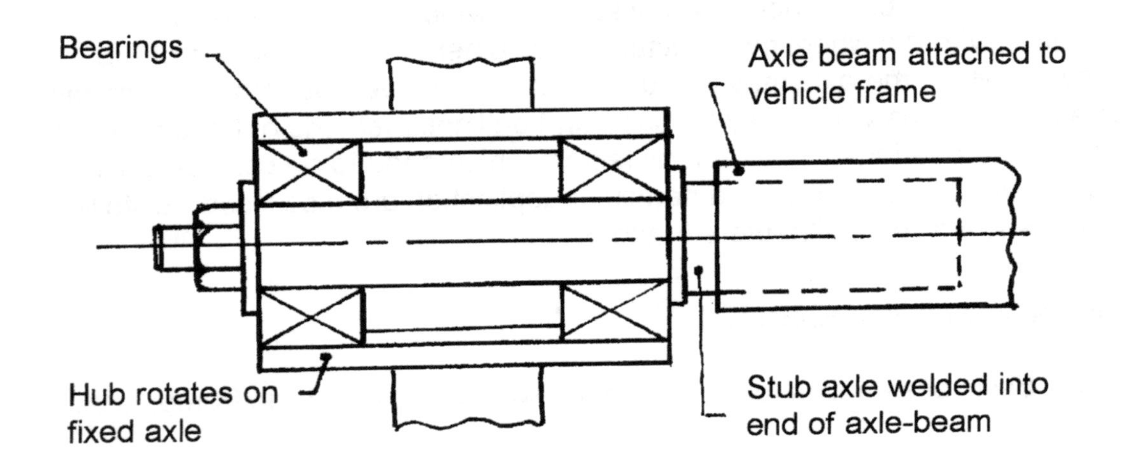


The bearings are mounted in the hub of the wheel and rotate with the wheel on a fixed axle. The inner end of the axle is attached to the vehicle frame. The wheel is pushed onto the axle and held in position by a retaining nut or pin.

**Advantages:**
simpler design of vehicle frame.

**Disadvantages:**
greater bending of axle and therefore larger axle size needed.

**Recommended Uses:**
for carts and trailers where commercial bearings (rolling element) or low-wearing bushes are used.

### 3. Rotating Axle supported at both ends

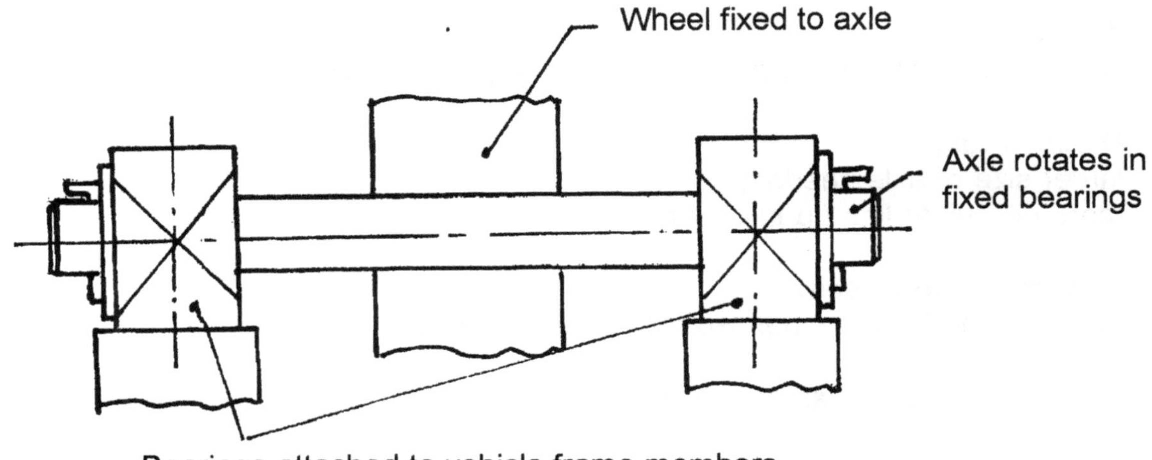


The wheel is fixed to the axle and the axle rotates in bearings fixed to the vehicle frame. The axle is supported in a bearing at each end.

**Advantages:**
less bending of axle and therefore smaller axle size;
lessens effect of wear in bearings because bearings are spaced further apart.

**Disadvantages:**
more complex design of vehicle frame needed and significantly increases width of vehicle;
more difficult to line up the bearings.

**Recommended Uses:**
for a wheelbarrow when wooden block bearings are used.

### 4. Rotating Half-Axles supported in fixed bearings (live axle)

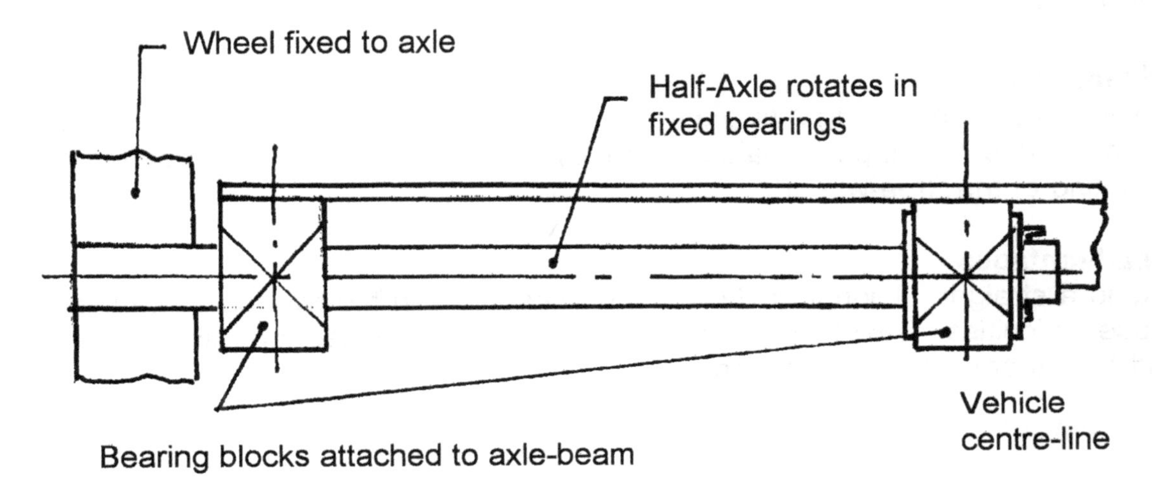

Each wheel is fixed to an axle which rotates in two bearings attached to the frame of the vehicle. One located at the outside of the frame and the other near the centre. Two half axles are therefore used and 4 bearing blocks.

**Advantages:**
significantly reduces effect of wear in bearings, since bearings are spaced well apart.

**Disadvantages:**
larger size of axle needed;
more difficult to line up the bearings.

**Recommended Uses:**
for carts and trailers when wooden bearings are used.

## 5. Rotating full-axle with free wheel

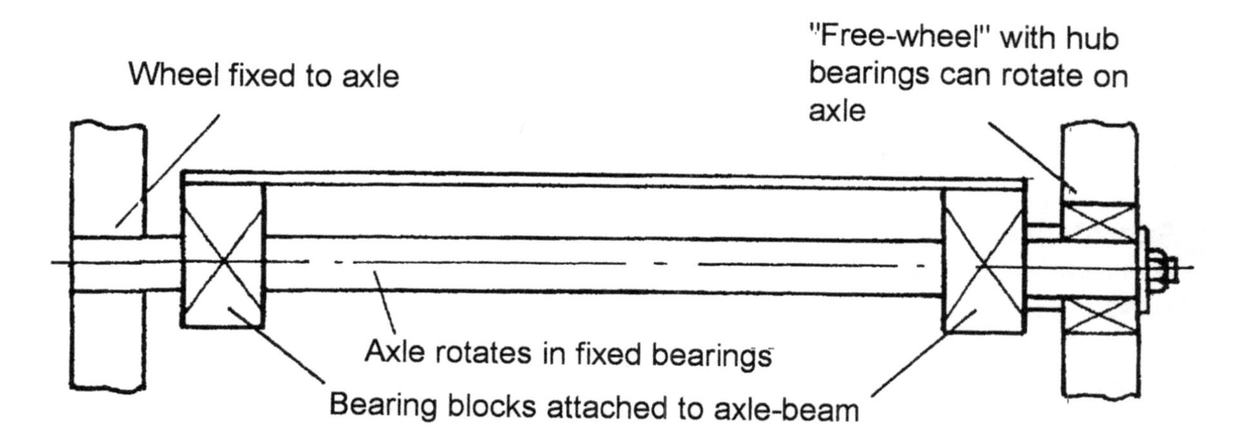


One wheel is fixed to a full length axle which rotates in two block bearings mounted in an axle-beam attached to the vehicle frame. The other wheel is free to rotate on the axle on bearings mounted in the wheel hub. The "free-wheel" should only rotate on the axle when the vehicles is not moving in a straight line and therefore wear of the hub-bearings should be quite low. Wooden on nylon bush bearings could therefore be used in the hub.

**Advantages:**
similar to 4, but mounting a single full-axle involves about half the work in mounting two half-axles, this is partly balanced out by the construction of a hub bearings and mounting of the "free-wheel".

**Disadvantages:**
finding a straight full length axle which rotates freely in the two fixed bearings may cause difficulties. Two types of wheels need to be produced, one fixed to the axle and the other with a hub bearing.

**Notes on Selection and Recommendations**

The main factors to be considered in selecting the type of bearing/axle assembly are as follows:

1. **Vehicle frame design**: the need to support the axle at both ends means that additional frame members have to be provided on the outsides of the wheels. These increase the weight, width and complexity of the frame. This type of assembly is therefore only recommended when a bicycle or motor-cycle axle is used (Type 1 assembly above), or the vehicle has a single wheel as on a wheelbarrow.

2. **Axle strength**: a rotating axle is more susceptible to fatigue failure because of reversed cycles of bending and therefore the axle size needs to be about 25% larger than for a fixed axle.

3. **Bearing wear**: wear in the bearings causes the wheel to tilt - the top of the wheel leans inwards and may rub against the vehicle body. The degree of tilt is determined by the amount of wear and the distance apart of the bearings. If the bearings are expected to wear it is desirable to place the bearings as far apart as possible. The **live-axle** assembly (Type 4) therefore has a significant advantage when the bearings are expected to experience substantial wear e.g. hardwood bearings.

4. **Alignment of bearings**: if the bearings are not properly lined-up the axle will `bind´ in the bearings causing excessive friction and wear. Fitting commercial bearings or machined bushes into an accurately machined hub (turned on a lathe) is the surest method of achieving proper alignment (i.e. Type 2 assembly). Separate bearings bolted to a frame (as in Type 4 assembly) are the most difficult to line-up and considerable care is needed in assembly to make sure the axle turns freely in the bearings.

## 4.2 TYPES OF BEARINGS

There are two basic types of rotating bearings:

1. **Rolling contact bearings** in which a rolling element (ball, roller or needle) is incorporated between the rotating and stationery parts of the bearing.

2. **Plain bearings** which involve sliding contact between the axle (or shaft) and bearing. This type of bearing may comprise a bush (or sleeve) fitted in a housing (hub), bearing shells fitted in a housing or a hole bored through a block of bearing material.

### 4.2.1 Rolling contact bearings

Since wheel bearings are subjected to combined radial and axial loads the most suitable types of bearings are:

- **Deep groove ball bearings:** these are the simplest and cheapest type. They are a single piece bearing comprising an inner race, which fits on the axle, and an outer race, which fits in the hub, separated by a ring of balls. For a **fixed axle** the **outer race** should be a **press fit** in the hub and the **inner race** a **push fit** on the axle. Although ball bearings cannot support as much axial load as taper-roller bearings they are quite adequate for the low-speed applications of non-motorised vehicles. Ball bearings can be fully shielded or sealed and prepacked with grease which is a significant advantage for operation in dirty conditions

- **Taper-roller bearings:** these have the greatest capacity for combined radial and axial loads and for this reason they are used on car and truck wheel mountings. The bearing is in two parts, an outer race which is fitted into the hub and an inner race with tapered rollers which is fitted on the axle. The bearing/axle assembly must allow careful axial adjustment of the bearings to ensure the two parts mate together with minimum clearance but without being over tight (pre-loaded). The bearings should be adequately lubricated and properly sealed to prevent dirt and sand getting into them.

### 4.2.2 Plain bearings

These may be further classified according to the method of lubrication into:

- **Full-film lubrication:** in which the speed of rotation and the supply of lubricant are sufficient to maintain a full-film of lubricant between the axle and bearing surfaces

- **Boundary lubricated bearings:** in which the supply of lubricant and/or the speed of rotation are inadequate to maintain a lubricating film and considerable sliding contact occurs between the surfaces of axle and bearing.

The bearings for low-speed vehicles will be of the second type. The relevant types for low-speed application are bush-type bearings and block-type bearings (e.g. hardwood).

The bearing material should be dissimilar from the axle material or the sliding surfaces should be hardened. Some of the common materials used are as follows:

- **Cast iron:** this is relatively cheap and performs reasonably well with a steel axle because of the graphite in its structure. However, because of the hardness of the cast-iron, the axle is best case-hardened to reduce wear of the axle surface. Adequate lubrication is essential

- **Bronze:** this is a more costly material but is widely used for bushes. The softer leaded-bronze is preferred and this can be used with a bright mild steel axle. If the harder phosphor bronze is used the axle should preferably be case-hardened. Adequate lubrication is essential

- **Case hardened steel:** this can be used with a case-hardened axle with adequate lubrication. It should be pointed out that ordinary mild steel bushes will not perform very well with a steel axle - they tend to wear rapidly and are prone to seizure. Because they are simple and cheap to make they are sometimes used, for instance on wheelbarrows, but they are really only suitable for low loads at low speeds

- **Non-metallic materials:** there are a wide variety of polymer bearing materials commercially available, some of which will out-perform the metallic materials mentioned above. However, high-performance polymer materials are not likely to be available in developing countries and the more commonly available materials such as nylon are unlikely to be adequate for wheel bearings

- **Wood:** hard wood performs quite satisfactorily as a bearing material providing that the bearing is properly designed and constructed. It is a cheap and readily available material which is easily worked. Little reliable information is at present available on wear rates, but if hardwood blocks are used in a live-axle arrangement, a bearing life of at least four to five years should be expected. With some designs wear can be counteracted to a certain extent and worn out bearings can be readily and cheaply replaced.

## 4.3 SELECTION OF BEARING TYPE

The selection will depend mainly on what bearings and machining facilities are available. The choices may be as follows:

### 1. Rolling contact bearings

The friction and wear in rolling contact bearings is much lower than in bush-type bearings but they need to be accurately mounted to perform satisfactorily and therefore it is preferable to machine both hub and axle on a lathe. However, it is possible to make a satisfactory assembly without the use of a lathe if suitable bright mild steel bar (shafting) is available for the axle, and care is taken to make a close fitting hub for the bearings (see Figure 6.4).

**Rolling contact, ball or roller, bearings are therefore the best choice if they are available and affordable.**

For low-speed vehicles the cheaper **deep-groove ball bearings** are adequate and these have other advantages:

i) they are supplied as a single piece and do not require precise axial adjustment. They are therefore simpler to assemble;

ii) they can be obtained fully sealed and pre-lubricated.

However, even though sealed bearing are used it is advisable to provide dust covers to minimise the ingress of sand and dirt into the hub. The bearings should be adequately sized to withstand the repeated impact loading on the wheel - at least Grade 2 bearings should be used i.e. for a 40mm diameter shaft 6208 bearings.

A typical hub assembly is shown in Figure 6.5. Two convenient sizes of bearings to use are:

1. For handcarts and cycle trailers with wheel loads up to 150kg:

   **6205 2RS** bearings - bore size 25mm<br>outer diameter 52mm; width 15mm<br>Hub can be made from 2" steel pipe, inside diameter 53mm.

2. For animal-drawn carts with wheel loads up to 600kg:

   **6208 2RS** bearings - bore size 40mm<br>outer diameter 80mm; width 18mm.

   Hub can be made from 3" steel pipe, inside diameter 80.8mm.

## 2. Bush-type bearings:

If ball or taper roller bearings are unavailable or unaffordable but a lathe is available then machined bushes could be used. Possible combinations are:

- cast-iron bushes with case-hardened axle;
- bronze bushes with case-hardened axle (a cold-rolled mild steel axle is just about acceptable for a soft bronze e.g. leaded bronze);
- case-hardened mild steel bushes with case-hardened axle.

The hub assembly would be similar to that shown in Figure 6.6. The hub should be made as long as possible and the bushes pressed in from each end. The length/diameter ratios for the bushes should be roughly 0.75 to 1.0. The initial diametral clearance in the bush should be about 0.001 to 0.002 x axle diameter (in mm).

A means for lubricating the bushes by oil or grease must be provided - the space between the bushes provides a useful reservoir for the lubricant. If oil is used this space may be filled with an oil absorbent material to retain the oil. If grease is used a shallow groove should be cut in the bore of the bush to spread the grease axially across the bush.

It is very important to prevent dirt and sand getting into the bearing as these abrasive materials will greatly accelerate wear of the bushes, and therefore effective dust covers are essential. Unsealed bush bearings are likely to last for only one or two thousand km.

If a lathe is not available it is possible to construct an effective bush-type bearing by winding 6mm diameter rod around a suitably sized former to produce a coil which will fit closely over the axle. However, the inner surface of the coil must be case-hardened to give a satisfactory wear life. If the grooves between the coils are packed with grease, this type of bush will remain efficiently lubricated over a long period of time. A bearing assembly using this type of bush is illustrated in Figure 6.7.

### 3. Hardwood block bearings:

If the use of rolling contact bearings is not feasible, hardwood bearings are a simpler and cheaper alternative to the bush bearings described above. Because of its relatively high wear rate, hardwood is recommended to be used as block bearings in a live-axle assembly rather than as bushes in a hub. In this form the effective life of the bearings is likely to be up to ten times greater than in a bush/hub assembly and possibly longer than that of the bushes described above.

As mentioned in Section 4.1 the advantage of the live-axle assembly is that the bearings can be spaced much further apart so that tilting of the wheel and axle caused by wear in the bearings is significantly reduced. Other advantages of fixed hardwood block bearings are that shrinkage of the wood is more easily catered for and the grain of the wood can be orientated to give the best bearing performance. The disadvantages are that the bearings are more difficult to line up and the outer bearing (closest to the wheel) is heavily loaded.

The problem of lining up the bearings can be partly overcome by fitting them into a channel beam formed by welding together two lengths of angle section. If the hardwood blocks are fitted tightly into the channel this will also reduce any tendency for the blocks to split. An assembly of this type is illustrated in Figure 6.8. The channel beam enables the axle assembly to be built up as an integral component which can readily be bolted to the frame of the vehicle in the same way as a stub-axle assembly.

Because the outer bearing supports most of the load, this bearing will wear more quickly than the inner bearing. The life of the assembly can therefore be significantly increased by interchanging the two bearings when the outer one becomes excessively worn. At the same time the clamping faces of the worn halves of the bearings can be planed to reduce the clearance in the bearings. (It may be necessary to fit a piece of packing under the inner bearing so that after the interchange the two bearings line up without binding on the axle).

The length/diameter ratio of the bearing blocks should be between 1 and 1.5 - the longer they are the more difficult they will be to line up. Because wood is not dimensionally stable a larger initial clearance is needed compared to other bearing materials - the bore should be machined to have an initial clearance on the diameter of about 0.02 to 0.03 x diameter. For example, for an axle diameter of 50mm the bore of the hardwood bearing should be 51 to 51.5mm.

Wood is stiffer along the grain than across the grain and evidence suggests that for the best bearing performance the wood grain should be end on to the axle. It also appears that this grain orientation gives better lubrication from an oil-soaked hardwood bearing. However, with this grain direction there may be some danger of the bearing splitting as indicated in the sketch and therefore as mentioned above, it is recommended that the bearing blocks are tightly fitted into a channel section beam.

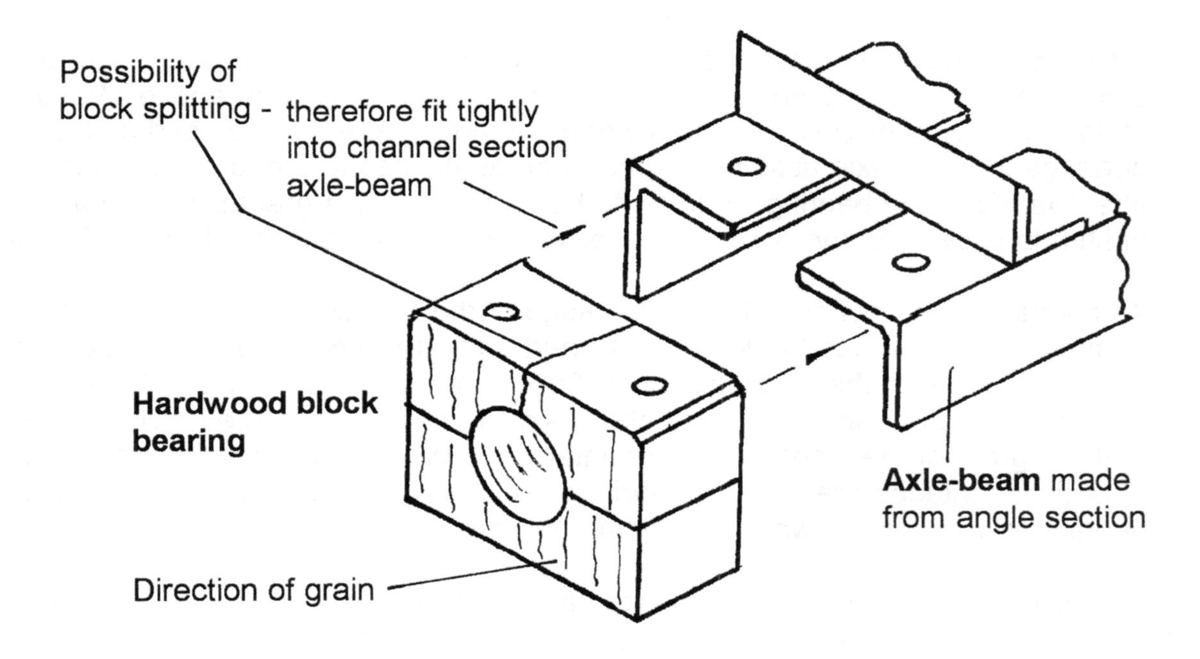

Grease or oil may be used for lubrication but oil appears to give the better performance. Because wood is a poor conducter of heat the bearings tend to run quite hot and grease forms a stiff gum which seems to increase friction. Also if sand or dirt gets into the bearing it may mix with the grease to form an abrasive paste which causes accelerated wear of the bearing.

A method of oil lubrication which is often used and which appears to be quite effective is to presoak the wooden bearings in oil. The bearings should be initially "boiled" in a container of oil for about 2 hours to drive out any moisture in the wood and then left to soak overnight in the cooling oil.

This form of lubrication should last for several months - this may be extended by drilling a number of holes in the **unloaded** half of the bearing and filling these with oil absorbent material soaked in oil. However, it is also advisable to provide an oil hole and filler plug so that further oil can be added in case the bearing runs dry and begins to squeek.

The construction of a simple, hand-operated device for accurately boring out hardwood bearing blocks is illustrated in Appendix 1.

## 4.4 AXLE DESIGN

The axle supports the wheel and transfers forces between the wheel and vehicle body. The radial and side loads on the wheel cause bending of the axle with maximum bending stresses on the top and bottom surfaces of the axle. The axle must be designed to be strong enough to withstand these bending stresses.

The shocks and impacts of running over stones, bumps, potholes etc. cause the wheel loads and resulting bending stresses in the axle to continually vary and therefore the axles have to be designed against fatigue failure. Fatigue failure is an accumulative form of damage in which the stress cycles (variations) build up until a crack starts in the material and the crack then grows until the material section breaks (fractures). Fatigue failures may therefore not occur until the vehicle has been operating for several years and axles need to be designed to last for an acceptable lifetime - for example at least 20,000km.

The critical factor for fatigue failure is the stress range which occurs at a point on the surface of the material i,e, the difference between the maximum and minimum stresses which occur at the point. Impact or shock loads are therefore very damaging and rigid-tyred wheels which cause higher impact loads will be more damaging than pneumatic tyres which cushion the impacts, so that axles need to be stronger for rigid-tyred wheels. Also live (rotating) axles are subjected to about double the stress range of stub (fixed) axles because the bending stresses at a point on the surface are reversed with each half revolution of the axle. The diameters of live axles therefore need to be about 25% greater than for stub axles in order to give twice the strength against fatigue failure.

The fatigue strength of materials is severely reduced by what are known as "stress concentrations". These are features which distrub the smoothness of a surface, for instance a corner, a change of section, a hole, a groove, a weld etc. For instance a sudden change in diameter (shoulder) of an axle reduces the fatigue strength of the axle by about 50% and the fatigue strength at the edge of a fillet weld will be only about 35% of the fatigue strength of the unwelded surface. Care therefore needs to be taken in designing and constructing axles to minimise these effects.

Guidelines on axle design are given in Figure 4.1. It is pointed out that bending of the axle should be minimised by keeping the centre-line of the wheel as close as possible to the critical section of the axle. Allowable loads for the wheel and axle are given for various axle sizes for typical layouts of stub and live axles. If the critical dimensions 'A' and 'B' are halved the allowable loads can be increased by about 25%.

## 4.5 AXIAL THRUST WASHERS

Vehicle wheels have to carry radial (vertical) loads to support the weight of the vehicle and load and also side (transverse) loads which result from cornering or from running over ruts, rocks, potholes etc. On low-speed vehicles cornering forces will be small and the main side loads will be the impacts from the latter effects. The axle assembly must therefore be designed to cope with these impact loads and also to prevent the wheel from moving from side to side. The axial location of the wheel or axle must be as accurate as possible (i.e. allowing only a very small axial movement) otherwise the impact loads will cause the wheel to hammer sideways against the locating members possibly causing damage. For instance if a split pin and washer is used and the pin is not accurately positioned, the wheel will hammer against the pin and may damage or break it.

Deep groove ball bearings and taper roller bearings can support both radial and side loads so that the problem in assembly is to axially locate the bearings on the axle. This is best done with a nut screwed onto a threaded end of the stub-axle, however, care must be taken not to overtighten the nut so that the bearings are preloaded. In the typical assembly shown in Figure 6.5 this is prevented by placing a spacer tube on the axle between the inner races of the two bearings so that when the nut is tightened it clamps the inner races against the spacer and prevents the tightening load from being transmitted through the bearing balls. With taper roller bearings some skill is needed to carefully tighten the nut so that it takes the slack out of the bearing without preloading it. With both types of bearings, after the nut has been tightened it needs to be prevented from working loose. This may be done with a split-pin (preferably with a castellated nut) or a lock-nut.

If plane bearings are used then additional axial (thrust) washers will be needed to locate the wheel or axle and to resist the side loads. These washers should have sufficient rubbing area to support the loads without rapid wear - the outside diameter should be at least 1.3 x the inner diameter. The washers also need to be fixed in such a way that rubbing occurs between the proper surfaces.

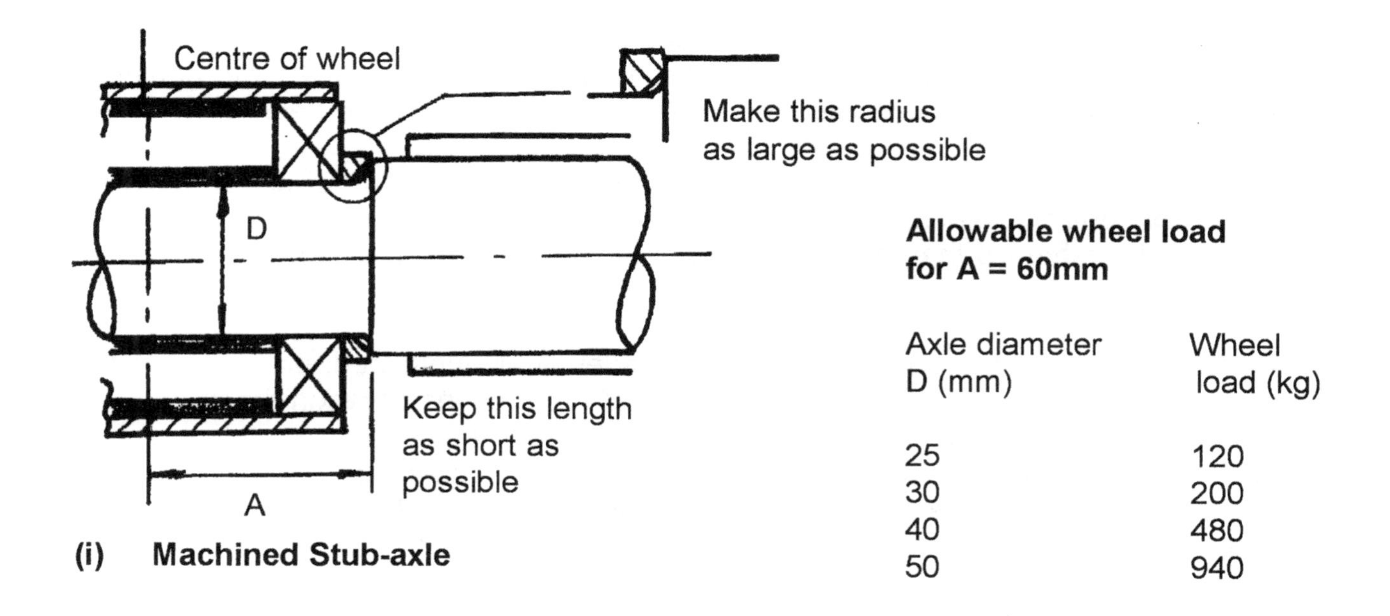


**Allowable wheel load for A = 60mm**

| Axle diameter D (mm) | Wheel load (kg) |
|---|---|
| 25 | 120 |
| 30 | 200 |
| 40 | 480 |
| 50 | 940 |

**(i) Machined Stub-axle**

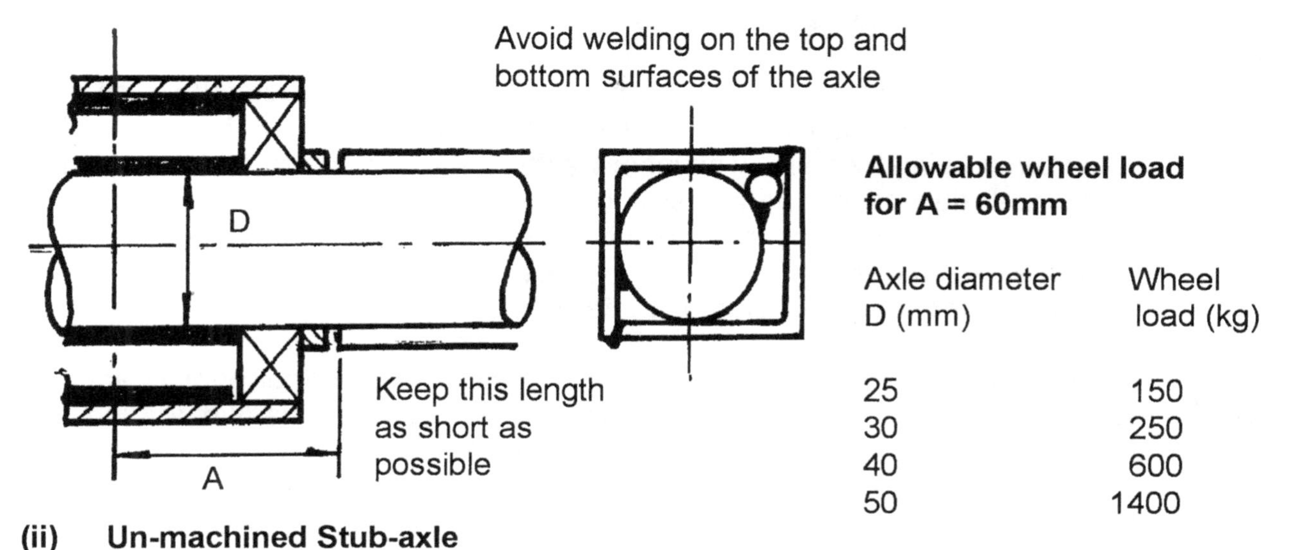


**Allowable wheel load for A = 60mm**

| Axle diameter D (mm) | Wheel load (kg) |
|---|---|
| 25 | 150 |
| 30 | 250 |
| 40 | 600 |
| 50 | 1400 |

**(ii) Un-machined Stub-axle**

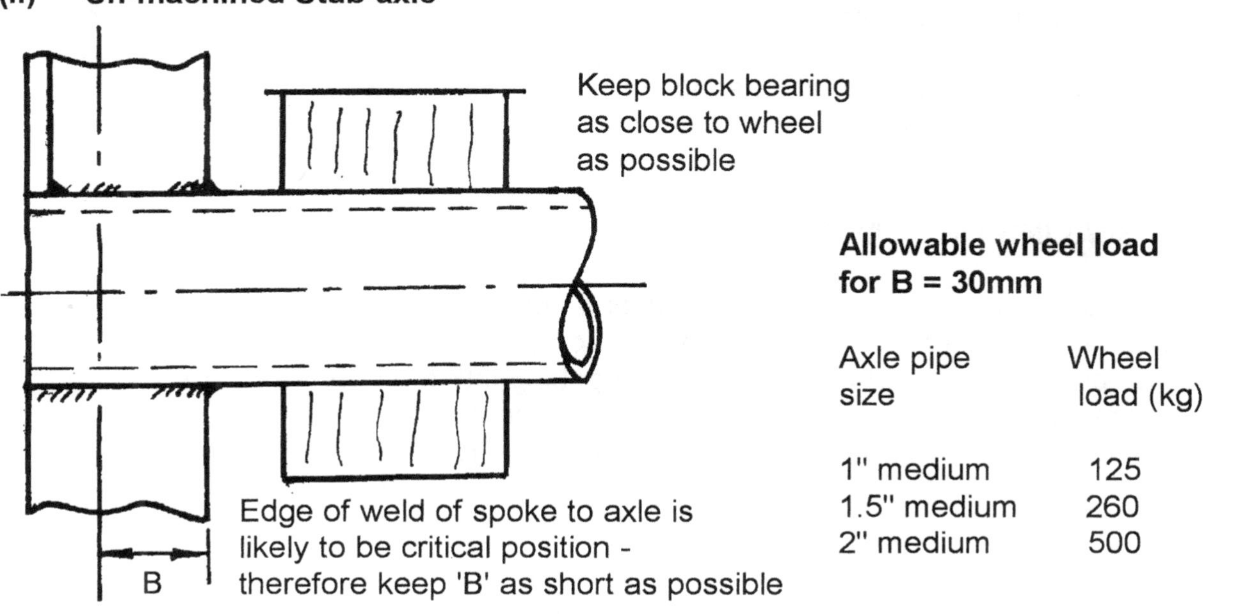


**Allowable wheel load for B = 30mm**

| Axle pipe size | Wheel load (kg) |
|---|---|
| 1" medium | 125 |
| 1.5" medium | 260 |
| 2" medium | 500 |

**(iii) Live-axle**

**Figure 4.1: GUIDELINES FOR AXLE DESIGN**

In general the thrust wash should be fixed to the axle so that it rubs against a matching face on the bearing. In some cases a pair of washers which rub against each other may be used, one fixed to the axle and the other to the bearing.

Washers may be cut or machined on a lathe from flat bar, or a flat spiral washer wound from 6mm diameter rod. At least two and preferably three rings are needed for a spiral washer so that the outside diameter of the washer will be 24 to 36mm greater than the inner diameter.

The rubbing faces of the bearing need to be adequately lubricated, preferably with grease. Spiral washers have the advantage that the grooves between the rings will retain the grease for long periods of time. For plain washers, shallow radial grooves may be cut from the inside to about 5mm from the outside diameter to provide better distribution of the grease. **If possible the rubbing faces should also be case hardened**.

Examples of the use of thrust washers in a stub axle assembly and live-axle assembly are shown in Figures 6.6 and 6.8 respectively.

In Figure 6.6 the thrust washers are fixed to the axle and the end faces of the bushes rotate against them. The outer washer is clamped against the end of the axle and is prevented from rotating by a pin. The inner washer has a slot which fits over a pin welded to the inner dust cover and is therefore prevented from rotating on the axle. The washers should be located so that the hub is just free to rotate with minimal axial clearance i.e. the axial movement of the hub between the two washers should be as small as possible without actually clamping the hub.

In Figure 6.8 the live-axle is axially located in both directions at the inner bearing which gives an easier assembly than splitting the location between the two bearings. Coil washers are fitted either side of the hardwood block bearing with the outer ends of the coil located in holes in the bearing to stop the washers rotating against the wood. Plain washers are fixed to the axle and rotate against the coil washers. The washers should be adjusted so that the axle is free to rotate with the minimum possible axial movement. It would also be possible to have plain washers rubbing directly against the hardwood bearing block but the hardwood would wear fairly rapidly and the lubrication of the washer would not be as good as with the coil washer.

### 4.6 SEALING OF BEARINGS

All bearings will perform more efficiently over a much longer period of time if abrasive materials such as dirt and sand are kept out of the bearing. This is particularly the case for plain bearings where sand and dirt getting into the bearing cause it to wear out much more quickly. Provision of effective seals or shields is therefore very important for achieving an acceptable operating life from the bearings.

Commercial seals, usually made from synthetic rubber, have a lip which fits tightly against the axle surface to prevent dust, dirt, etc. getting into the bearing and lubricant from leaking from the bearing. They need to be precision-mounted to seat on a ground axle surface. They are unlikely to be available or suitable for the wheel assemblies presented in this manual.

Simpler forms of seals for excluding dirt etc. consist of felt or leather rings fitted into a groove in the hub or bearing housing and rubbing on the surface of the axle. It may be possible to construct a simple form of this type of seal using leather or greased (or waxed) twine. Two ideas are outlined in Figure 4.2 for sealing the inner end of a hub on a stub-axle - the outer end of the hub can be completely sealed by fitting an end cap tightly over the hub to enclose the end of the axle.

Rubbing seals of the above form create considerable friction and also tend to wear. An alternative is to use a shield arrangement which involves no rubbing contact and therefore no friction and wear. The shield attempts to obstruct the path of dirt, sand etc. into the bearing area. It is not as effective as a seal but if properly construction should be quite adequate.

The hub assemblies shown in Figures 6.5 and 6.6 have shields fitted at their inner ends. These are tightly fitted over the axle and should have enough clearance around the hub so that the axle can rotate without scraping on the shield.

More clearance is needed in the case of the bush-type bearings to allow for some wear of the bushes. The space between the shield and hub may be sealed with grease. This may be reinforced by a length of scrap inner tube tied tightly around the shield to form a sealing lip against the hub.

The design of shield type seals is illustrated in Figure 4.3.

A shield is also an effective method for restricting the entry of sand and dirt into a hardwood block bearing in a live axle assembly. The shield cover may be made as a single part or in two parts bolted together - the latter may be easier to assemble. It should be screwed tightly against the end face of the hardwood block to prevent any sand or dust getting in at this joint - if necessary, a sealing washer (gasket) can be made from scrap inner tube to seal any gaps between the cover and the wood. A gap of 1 to 2mm needs to be left between the shield cover and the axle to allow for wear in the bearing.

An assembly of this type is shown in the Figure 4.4. A washer is welded to the axle to shield the bearing. If a thrust washer is needed to locate the axle against the bearing then either the washer shown may bear directly against the hardwood block or preferably a coil washer may be inserted as indicated by the dotted outline.

Figure 6.8 shows a live-axle assembly with shields fitted to hardwood block bearings. The axle is located at the inner bearing by coil thrust washers.

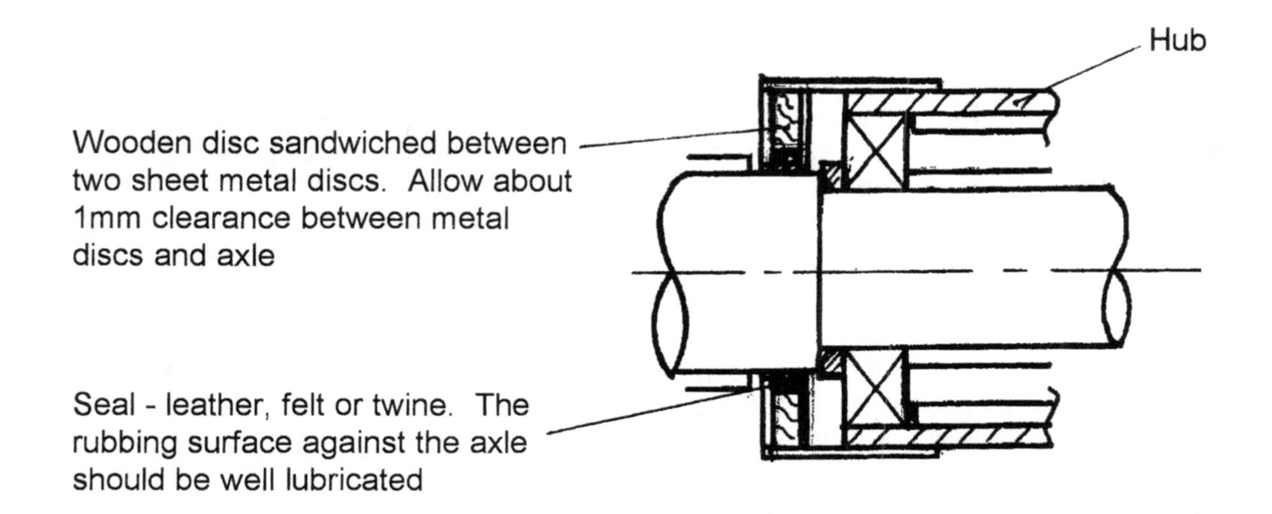


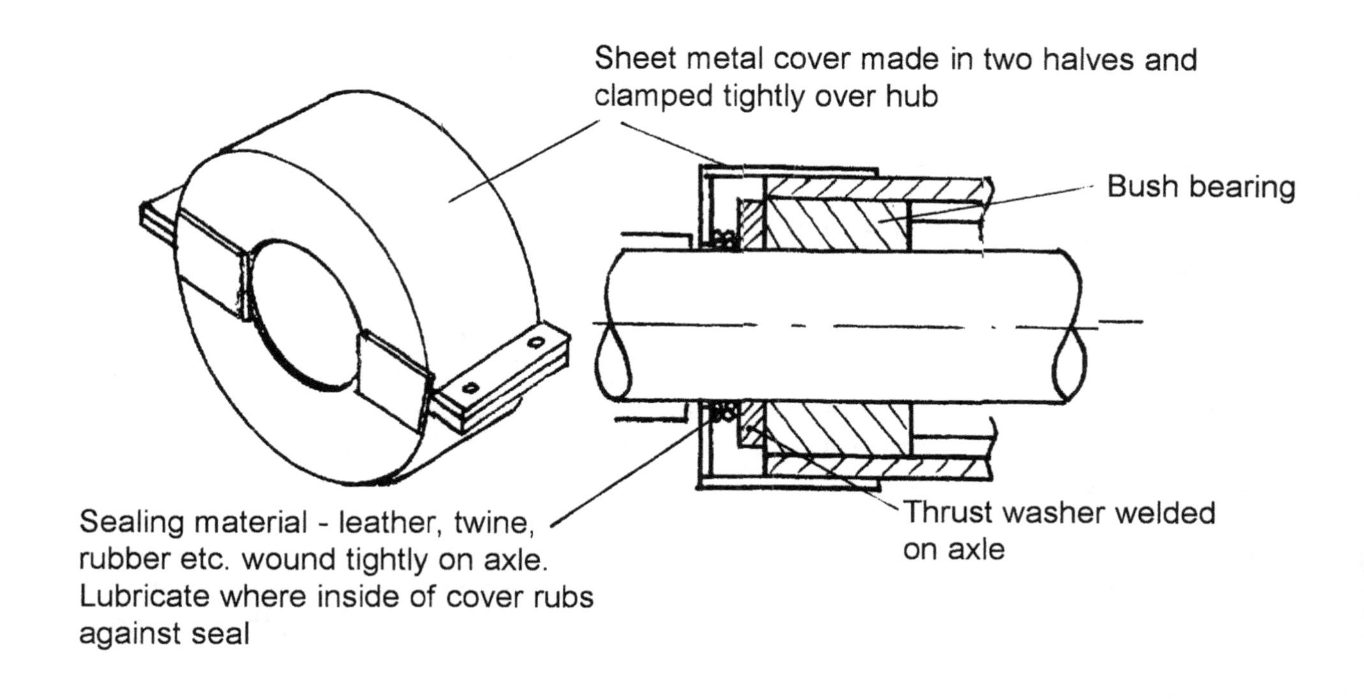


**Figure 4.2: TWO TYPES OF BEARING SEAL FOR A HUB ASSEMBLY**

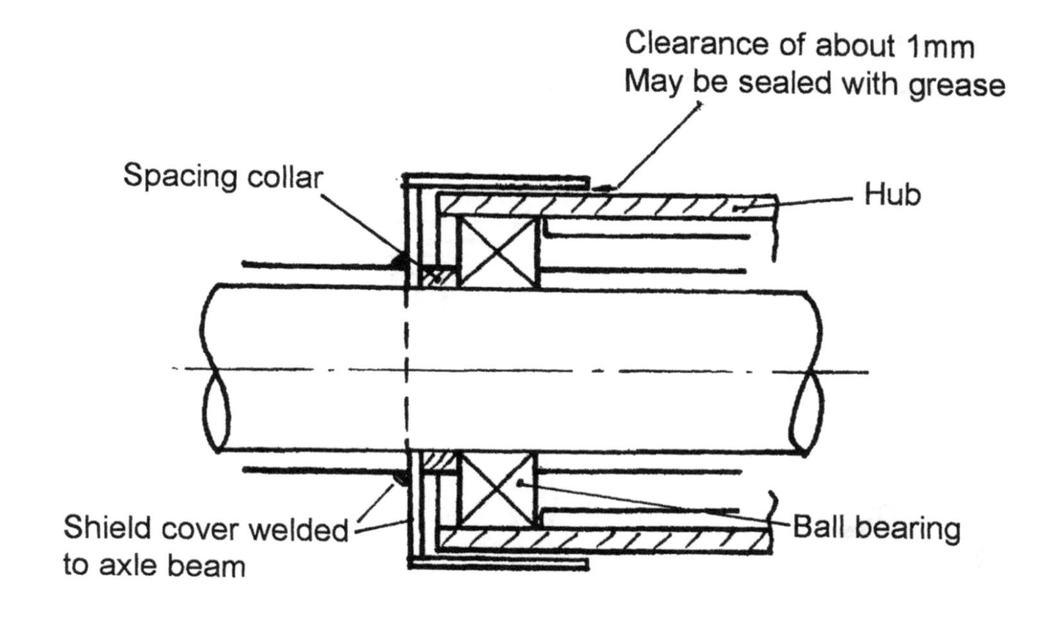


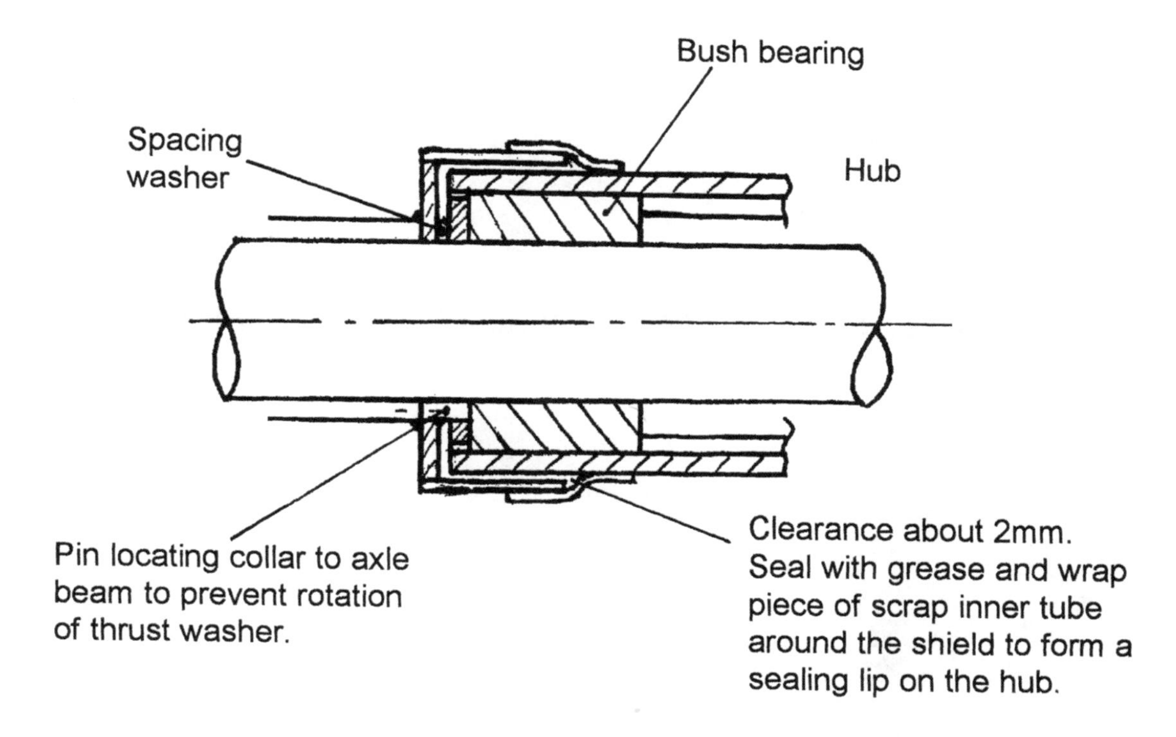


**Figure 4.3: SHIELDS USED FOR SEALING A HUB ASSEMBLY**

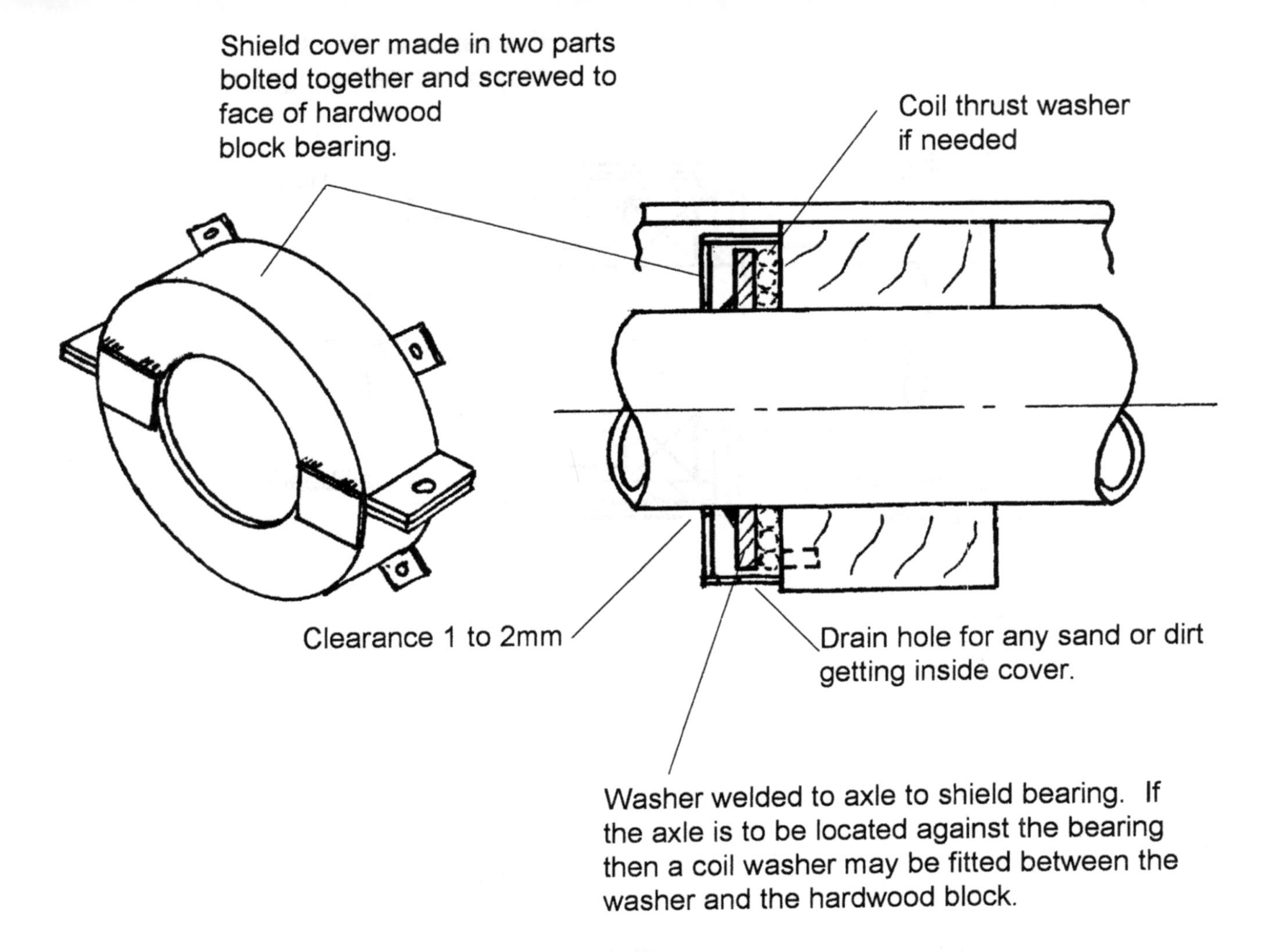


**Figure 4.4: A SHIELD USED TO SEAL A HARDWOOD BLOCK BEARING**

# CHAPTER 5. MANUFACTURE OF WHEELS TO TAKE BICYCLE AND MOTORCYCLE TYRES

This chapter gives details of the manufacture of wheels with angle-section rims which can be fitted with bicycle, motorcycle or solid rubber tyres. The wheels have spokes which may be angle section or round bar. Various hub-axle assemblies may be used with the wheels - the bicycle-type wheels may be fitted with standard bicycle wheel axles which need to be supported on both sides of the wheel or any of the wheels may have hub and stub-axle assemblies or live axles as outlined in Chapter 4.

The following details are included:

**Section 5.1** gives details of the design of bicycle-type wheels which can be fitted with 28 x 1 1/2" or 28 x 1 3/4" bicycle tyres. The design can readily be adapted to take other sizes of bicycle tyres or solid rubber tyres. The wheels are designed for loads up to 100kg and are intended for use on bicycle trailers or handcarts. Details of two designs are given with different spoke arrangements.

**Section 5.2** gives details of the construction of a hub to take a standard bicycle axle. Construction of other hub axle assemblies are briefly outlined.

**Section 5.3** gives step-by-step instructions for construction of the bicycle-type wheels. The steps are exactly the same for the manufacture of other spoked wheels with angle section rims.

**Section 5.4** describes the manufacture of a wheel with an angle section rim which can be fitted with a motorcycle tyre. The wheel is designed for loads up to 200kg and is intended for use on heavy duty handcarts or light donkey carts.

## 5.1 DESIGN OF WHEELS TO TAKE BICYCLE TYRES

These wheels have rims bent from 25 x 25 x 3 angle section. The designs shown have rim dimensions to take 28 x 1 1/2" or 28 x 1 3/4" bicycle tyres, but the rim sizes can be readily changed to suit other tyre sizes.

Two designs are shown:

**Figure 5.1** has spokes cut from the same angle as the rim. This is the most robust of the designs but is also the heaviest and the most difficult to construct - also the overlap of the spokes onto the rim prevents the use of rim brakes on the wheel. The welding of the spokes to the rim tends to cause more sideways distortion (wobble) of the rim than the other designs, especially if the spokes are offset to the ends of the hub as shown in the Figure. The offset is to provide more space to weld all around the angle at the spoke to hub joint, it does not make the wheel stronger. Therefore, if there are problems with too much sideways wobble of the wheel the spokes can be moved more towards the centre of the hub. The weight of the wheel without tyre and tube is about 4.5kg.

**Figure 5.2** has spokes cut from 8 diameter round bar (reinforcing bar). The spokes are used in pairs to give adequate strength against side loads on the wheel. This wheel is the simplest to make but good penetration welds are needed at each end of the spoke to provide adequate strength for the wheel. The weight of the wheel without tyre and tube is about 4kg. The wheel can be fitted with rim brakes and could therefore be used as a strengthened wheel on a bicycle.

Although the wheels are significantly heavier than a standard bicycle wheel they are also considerably more robust and more suited to operation on rough earth tracks where large side loads caused by hitting rocks or pot-holes would be likely to cause bicycle wheels to buckle.

**Note:** that it is very important that the bicycle tyres should fit as tightly as possible on the rims to minimise the risk of the tyres coming off during operation. The rims should therefore be made as large as possible whilst still allowing the tyre to be levered on without being damaged.

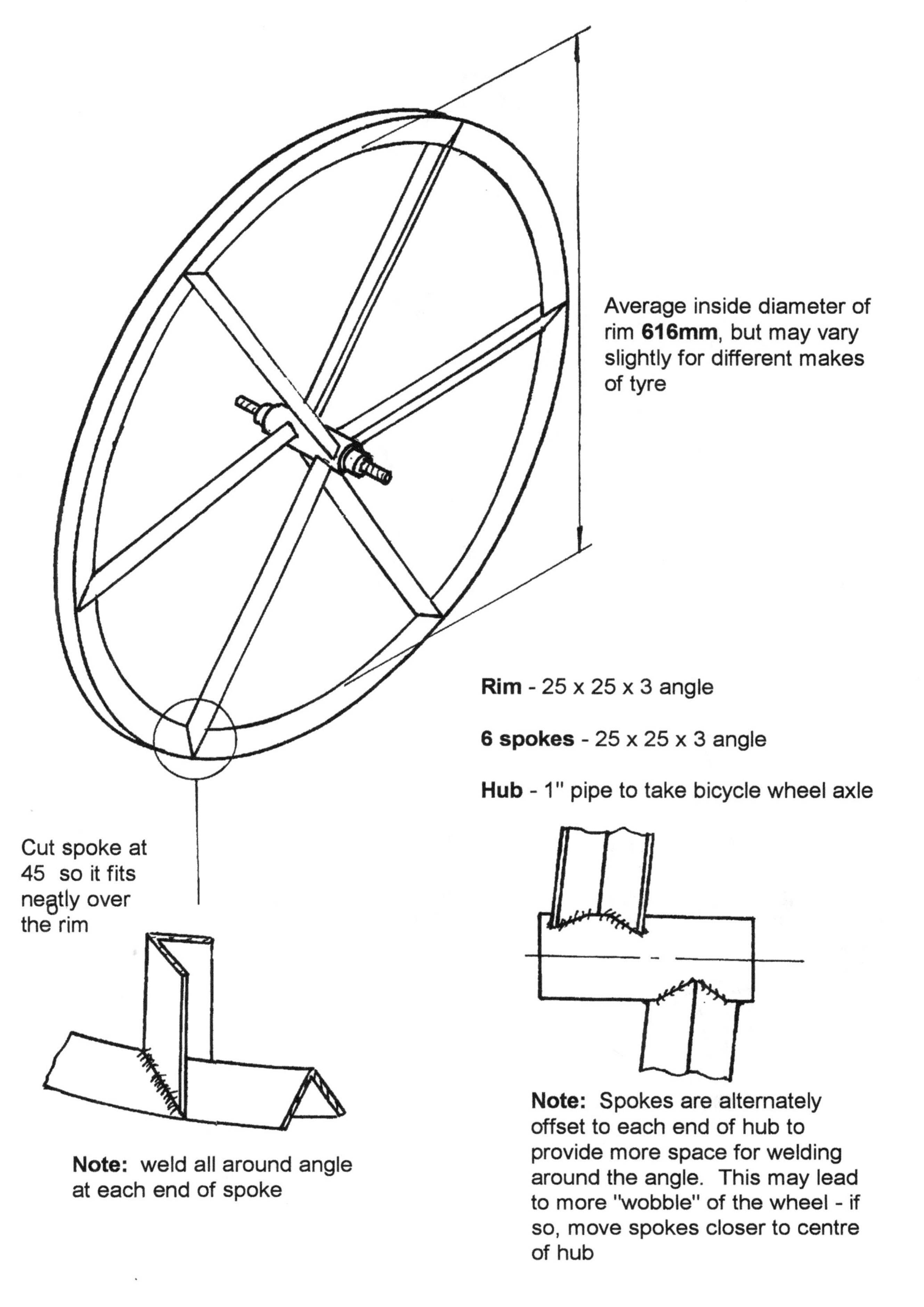


**Figure 5.1: DESIGN OF WHEEL TO TAKE 28" BICYCLE TYRE**

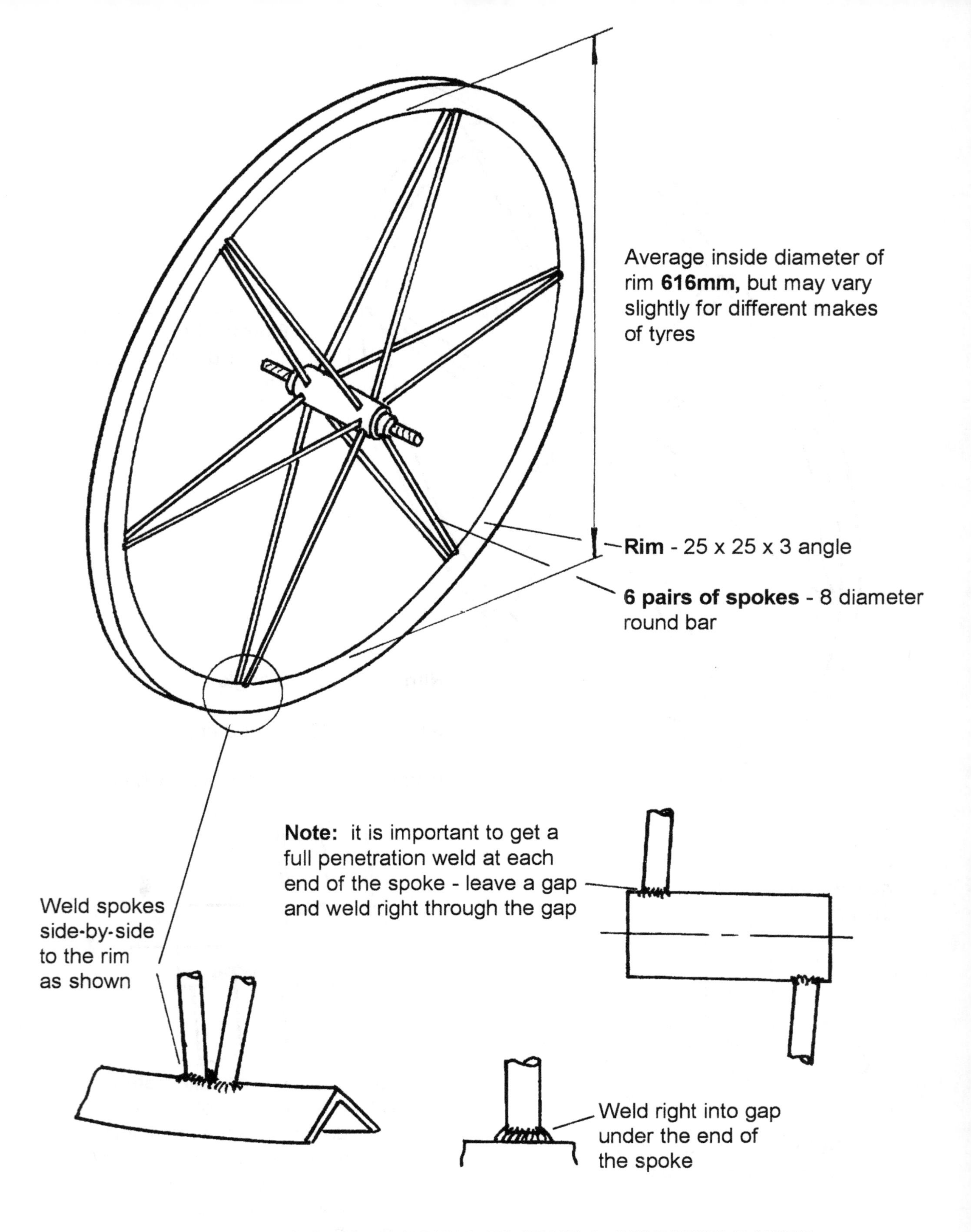


**Figure 5.2: DESIGN OF A WHEEL TO TAKE A 28" BICYCLE TYRE**

## 5.2 CONSTRUCTION OF HUB AXLE ASSEMBLY

Three types of wheel axle assembly may be used with the wheels:

1. **Bicycle wheel axle:** this design is shown in Figure 5.3. It uses a 3/8" axle together with bearing balls and cups - these parts can be obtained from a bicycle spare parts supplier. The bearing cups are fitted into each end of a hub machined from 1" medium-wall pipe and are located in position by a sleeve cut from 3/4" pipe and plug welded inside the hub. The bore of the hub must be machined on a lathe, or the ends opened out by pressing or hammering in a punch tool of the required diameter. The cups must be a tight fit in the hub so that they have to be pressed or tapped into place. Sealing washers should be used to keep sand and dirt out of the bearings. Although this design is efficient, if the wheels are continually used to carry heavy loads on rough roads the bearing cups may be damaged causing the bearing to collapse. However, the assembly can easily be repaired at relatively low cost.

2. **Stub axle assembly:** a suitable design is shown in the next chapter in Figure 6.11. It uses a 25mm diameter axle and two deep groove ball bearings types 6205 2RS. The bearings are fitted into a hub fabricated from 2" medium wall tube.

   Although this is a heavy duty axle assembly for a bicycle type wheel it is the most convenient to construct. If a 20mm diameter axle is used the bearings type 6204 2RS have an outside diameter of 47mm which does not match with a standard pipe size. However, if a lathe is available it may be possible to machine a hub to take these bearings.

   Bush bearings of the type described in Chapter 4 may be used instead of ball bearings provided they are properly lubricated and sealed.

3. **Live axle:** an assembly similar to that shown in Figure 6.8 may be used, using two half axles cut from 1" medium wall pipe supported in hardwood block bearings. In this case the spokes of the wheel will be welded directly to the axle.

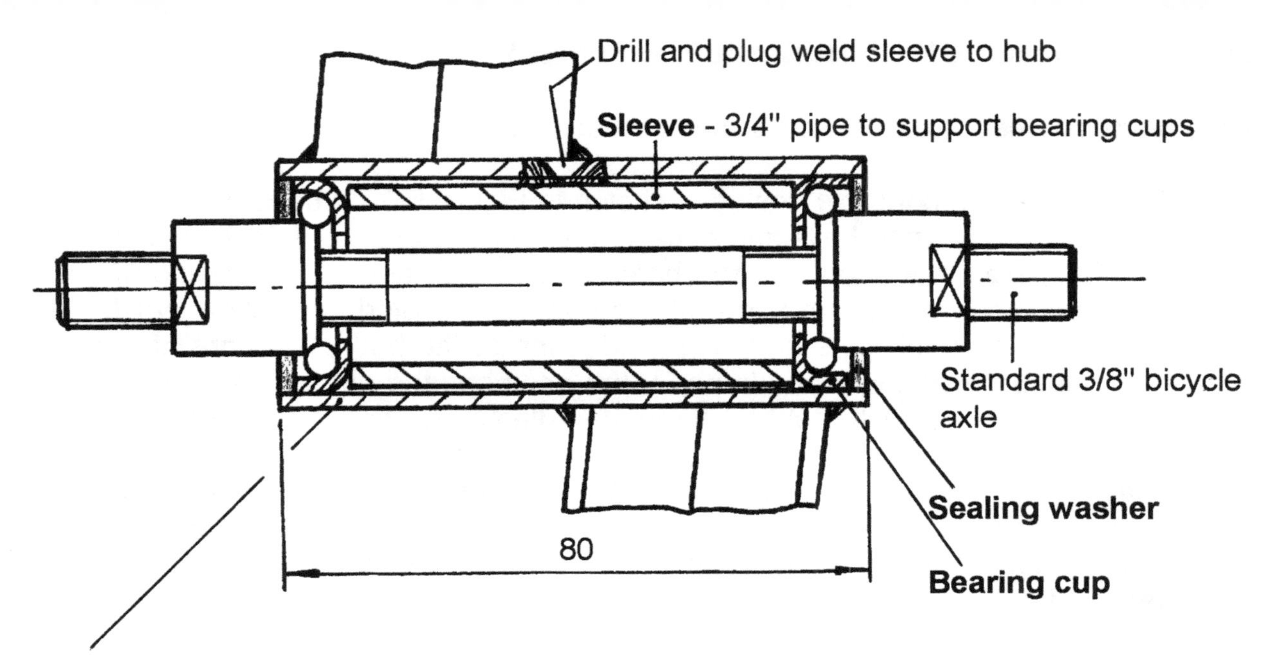


**Hub:** 1" medium wall pipe, bore out so that bearing cups are a tight push fit in the ends.

**Or:** machine a punch of the shape shown and hammer it into the ends of the pipe to open out the bore so that the bearing cups can be pressed or tapped into position.

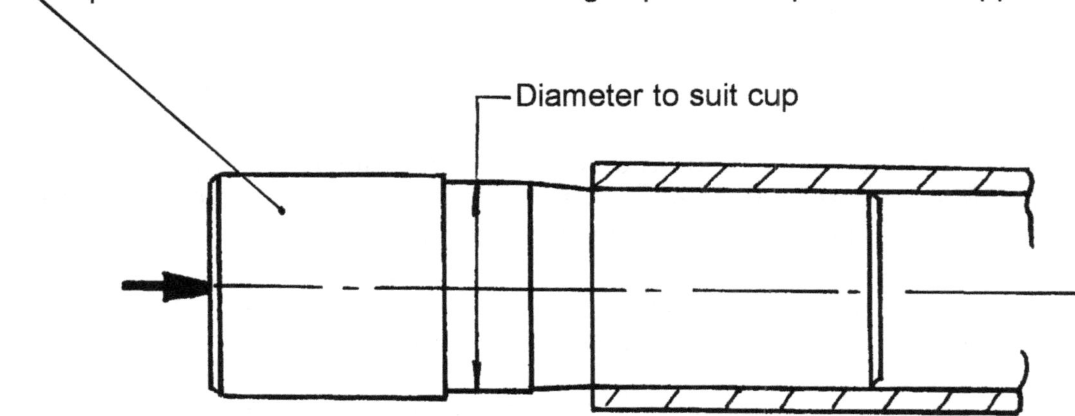


**Steps in construction of hub:**

1. Cut 80mm length of 1" pipe for hub - 34mm outside diameter, 27mm inside diameter.

2. Machine bore of pipe to suit diameter of bearing cups **or** use a machined punch to open up the ends of the pipe to fit the cups. The cups should be a tight fit in the hub.

3. Cut a length of 3/4" water pipe for sleeve, drill a hole in the hub and plug-weld the sleeve in position.

4. Fit cups making sure they line up accurately so that axle rotates smoothly.

**Figure 5.3: A HUB ASSEMBLY USING A STANDARD BICYCLE WHEEL AXLE**

## 5.3 CONSTRUCTION OF BICYCLE-TYPE WHEELS

This section gives step-by-step instructions for the construction of a wheel to take a 28" bicycle tyre - the wheel designs are shown in Figures 5.1 and 5.2.

The detailed instructions are followed by a set of photographs and notes which illustrate and summarise the steps.

**Note:** that other wheel designs with an angle-section rim can be made in exactly the same way except the setting of the adjusting screw on the bender needs to be changed to suit the size of rim required.

### Step 1: Forming the Wheel Rim

1.1 Cut length of 25 x 25 x 3 angle of 2350mm (mean diameter of rim = 635mm therefore mean circumference = 635$\pi$ = 1994mm. Add 356mm allowance for ends).

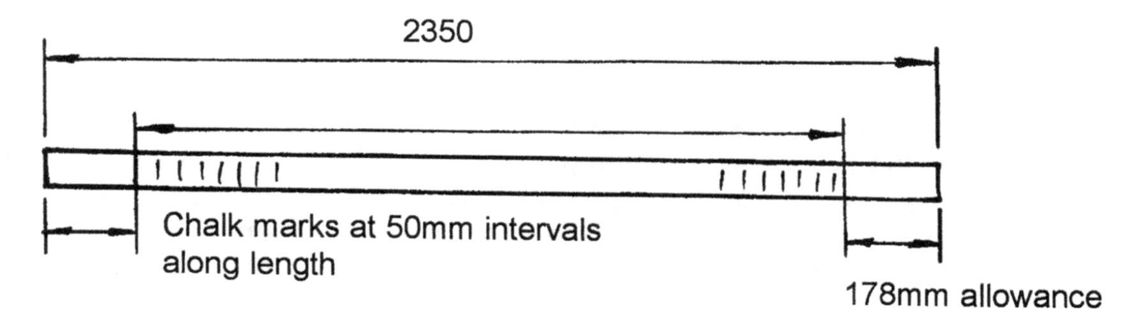


1.2 Make saw cuts across angle 178mm from each end.

1.3 Mark out length between saw cuts in 50mm intervals with chalk.

1.4 Place lower rollers in **Outer** position and set up the **'V' tool** in the tool holder.

1.5 Rest a **straight** piece of angle section across the lower rollers and **"Zero"** the bending device (see Section 2.3 and photographs in Section 6.3).

1.6 Screw down the **adjustable stop** by 10 turns (25mm) and tighten the lock-nut.

**Note:** The actual setting depends upon the "spring back" of the angle section after bending and will vary from batch to batch of angle. Some initial experimentation with the stop adjustment may therefore be needed to achieve the required wheel diameter. Once the correct setting is established the stop can be locked in position with the lock-nut for bending of subsequent rims.

1.7 An extension lever of about 1 to 1.5m of 1 1/2" water pipe should be placed in the lever socket.

1.8 With the leading end of the angle just resting on the furthest roller bend the rim by forcing the lever arm down onto the stop. Feed the rim through the bender in the steps shown by the chalk marks and at **each step** bend the rim.

**Note:** the lever arm must touch the stop at each bend but do not press down excessively hard on the stop.

1.9 Towards the end of the bending process displace the rim section to the right so that it passes along the outside of the bender where the bottoms of the angle section columns have been cut away.

1.10 When the rim is fully formed remove the lever arm and lift the rim out of the machine.

1.11 Cut one straight piece off one end of the rim at the saw mark - make sure the cut is square.

**Step 2: Check the size of the rim**

**Note:** The tyre must be as tight a fit as possible on the rim to avoid later problems with the tyre coming off the rim during service. All tyres are not exactly the same size, so slight adjustments may need to be made to the rim diameter. In particular, tyre sizes tend to vary slightly between different makes. Therefore, for each new type or batch of tyres a trial rim should be made up to check the fit of the tyre.

2.1 Set the positions of the **seven guide stops** of the assembly jig using the gauge. The gauge should be set initially as follows:

Gauge length = Radius of inside of rim - 1/2 of centre post diameter

Example: if centre post diameter is 20mm - gauge setting
= 308-10
= 298mm

2.2 Place the rim in the assembly jig and tighten the bolts in the sequence shown in Figure 5.4.

2.3 At the rim joint, cut off the straight overlapping end leaving a gap of about 2mm for welding between the two ends. Tack-weld the joint.

2.4 Remove the rim from the jig and try the tyre for size.

- if the tyre does not fit tightly enough, cut the tack weld and weld in a short extra length of angle;
- if the tyre is too tight, cut a short length out of the rim.

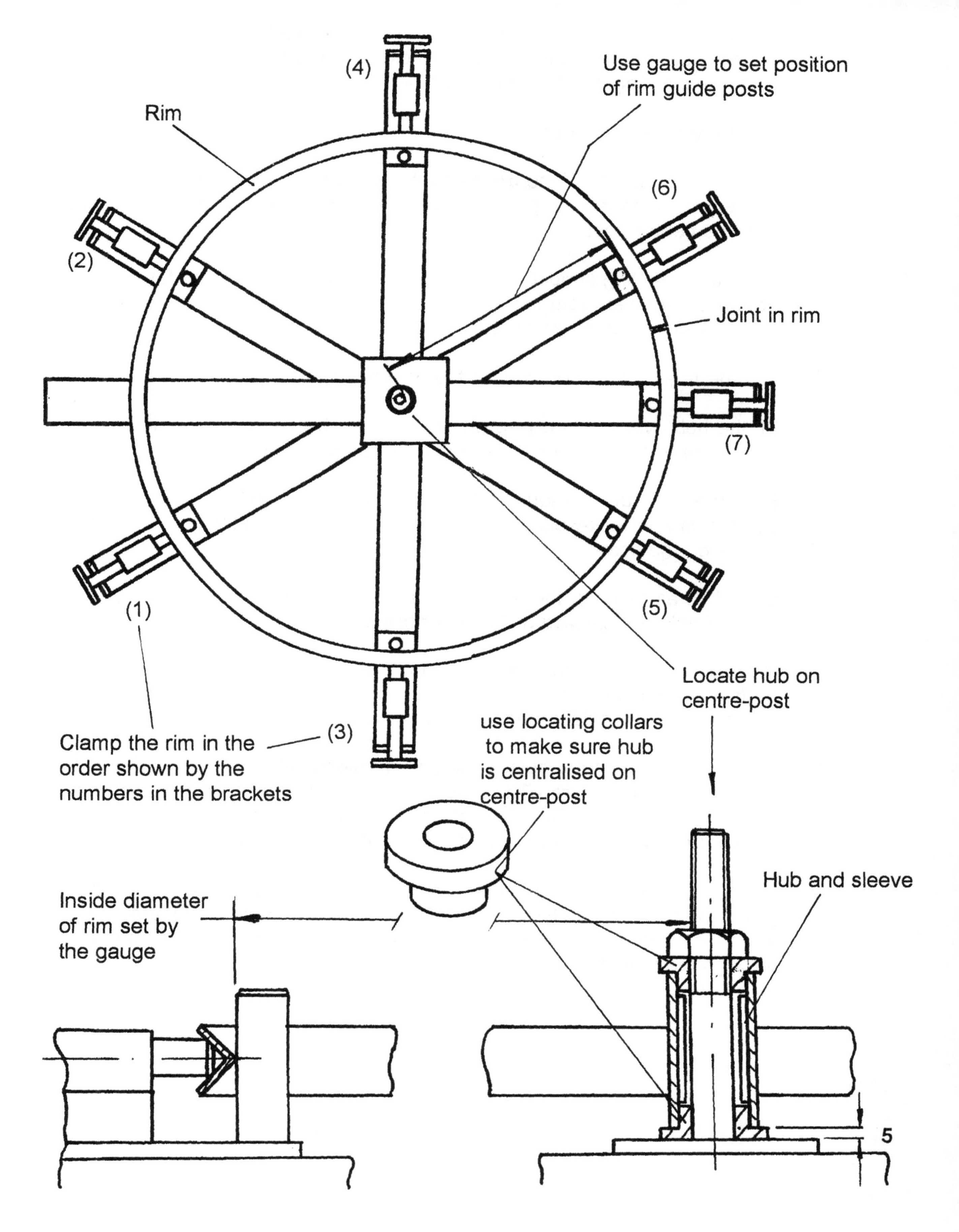


**Figure 5.4: SETTING UP THE WHEEL IN THE ASSEMBLY JIG**

2.5 When a satisfactory fit of the tyre has been achieved note the length added or removed from the rim and for the next rim, add or subtract this length from the cut length (2350mm) of the rim.

**Step 3: Assemble the wheel**

3.1 Adjust the positions of the **guide stops** as required to suit the new diameter of the rim, using the gauge provided to ensure the rim remains concentric with the centre-post.

3.2 The gauge setting should be increased or decreased by 1mm for each 6mm added or removed from the rim. Replace and lock the rim in position in the jig and weld as much as possible of the joint.

3.3 Cut the required hub or axle for the bearings to be used - see Section 5.2.

3.4 Fit the hub over the **centre-post** of the assembly jig, use the collars to locate the hub centrally on the post and clamp it in position.

3.5 Measure and cut the required lengths of angle or round section for the spokes, fit and tack-weld them in position. The spokes should be located alongside the **stops**.

3.6 Complete as much of the welding as possible whilst the wheel is in the jig - this will help to minimise distortion of the wheel.

3.7 Remove the wheel from the jig, complete the welding and clean up. Pay particular attention at the inside of the rim section to file or grind the weld smooth to prevent damage to the inner tube.

3.8 Drill a 8.5mm diameter hole in the rim, **diametrically** opposite to the joint, for the valve of the tyre. The hole should be drilled through the centre of the 'V' of the angle. Chamfer (take the sharp corner off) the edge of the valve hole inside the rim to avoid cutting of the inner tube.

3.9 It is best to wrap a strip of inner tube around the inside 'V' of the angle to give protection for the tyre inner tube and also to fit a small square of scrap inner tube over the valve stem to give protection where the valve passes through the hole in the rim.

## CONSTRUCTION OF WHEEL TO TAKE 28" BICYCLE TYRE

The photographs and notes illustrate and summarise the step-by-step instructions for making a bicycle-type wheel.

### 1. Set up the bender

- Fit 'V'-tool and lower rollers in outer position
- Rest **straight** length of rim on rollers, apply slight downward pressure on lever arm and adjust screw until it just touches stop on lever arm
- Screw adjusting screw down **10 turns** and lock screw in position.

### 2. Bend the rim

- Start with end of rim just resting on furthest roller
- Bend the rim by forcing the lever down on to the stop
- Push rim through to next chalk mark and bend rim again.
- Keep feeding rim in steps shown by chalk marks and bend rim at each step. **Make sure** that lever touches stop at each bend.

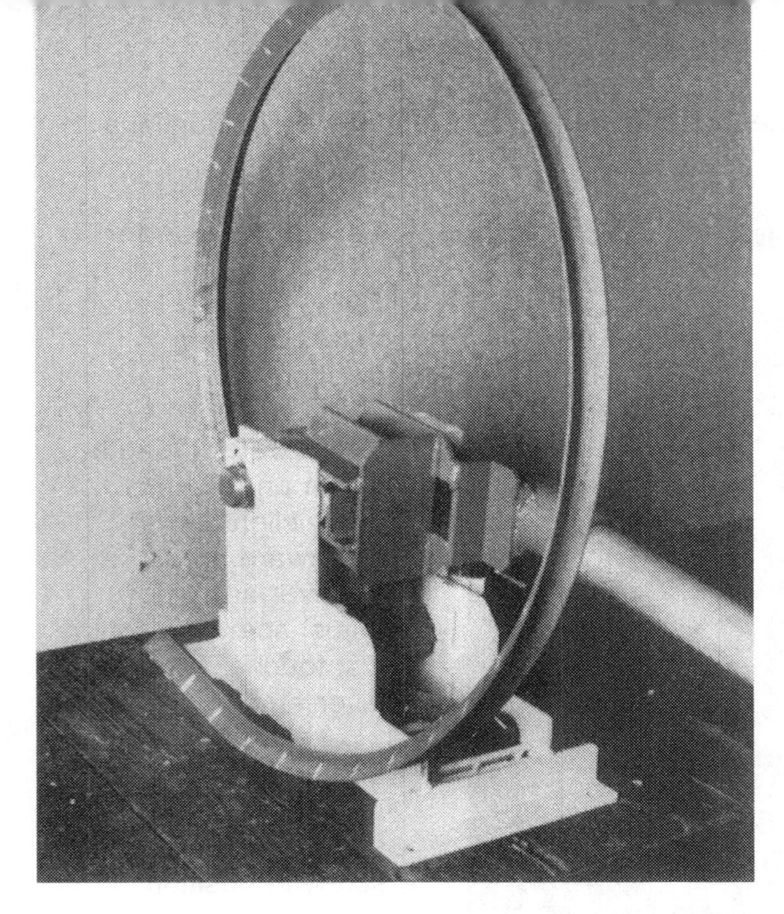

### 3. Fit rim in assembly jig

- complete bending of the rim. Note the overlap of the straight ends of the rim;
- remove the rim from the bender;
- cut one straight end off at the saw mark.

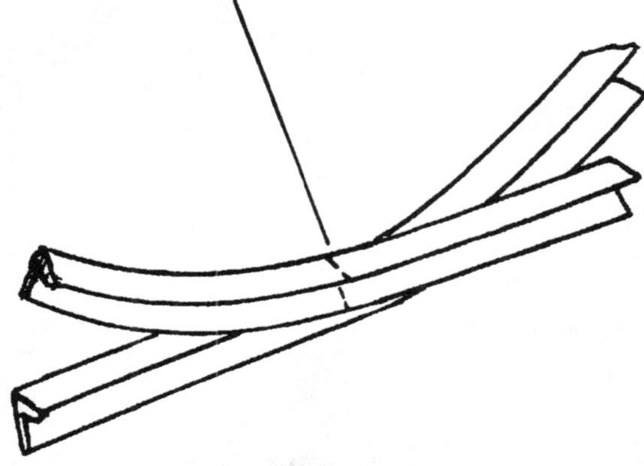

- use the gauge to set the rim clamps in the correct position for the rim size on the assembly jig;
- fit the rim in the jig and tighten the clamps in the sequence shown in Figure 5.4;
- at the overlap of the ends of the rim cut off the straight section to leave a gap of about 2mm for welding.

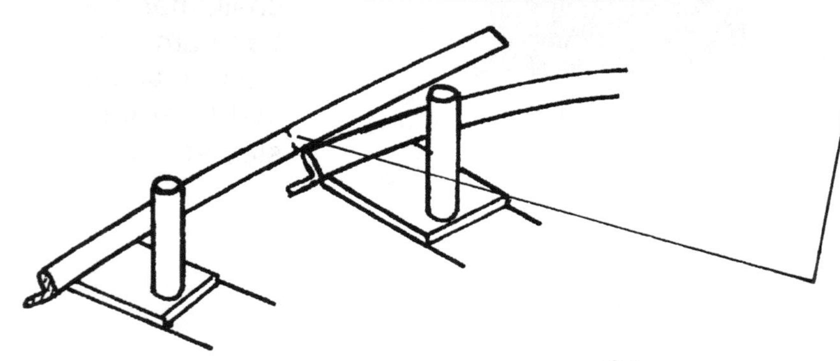

- weld up rim joint;
- fit hub over centre post. Use locating collars to make sure hub is central and square on post;
- clamp hub in position;

- measure length of spokes and cut 12 pieces of 8 diameter rod **or** 6 pieces of 25 x 25 x 3 angle;
- hold spokes in position against the stops and tack-weld them to rim and hub;
- remove the wheel from the jig and complete the welding. Try to balance the welds to keep distortion of the rim as small as possible;
- grind or file the rim joint weld smooth on the inside of the rim and drill a hole for the valve opposite to the joint.

**Completed wheel** fitted with bicycle tyre and tube.

## 5.4 MANUFACTURE OF A WHEEL TO TAKE A MOTORCYCLE TYRE

**Figure 5.5:** shows the design of a wheel to take a **17"** motorcycle tyre. This is a simpler design to make than the detachable bead wheel described in Section 6.4. The details of the design and manufacture of the wheel are outlined below.

**Rim:**

- This is bent from **40 x 40 x 3** angle section
- The inside diameter of the rim is 400mm and the outside diameter 456mm
- The diameter of the neutral axis of the rim (the axis about which the angle bends) is 422mm so that the length of the rim is **1325mm**
- Add on 350mm for the end allowances so that the length of angle to be cut is **1675mm**.

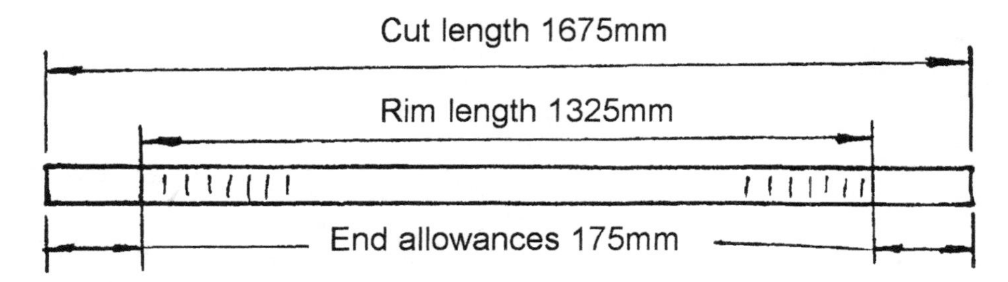


**Bending the rim:**

- Set the lower rollers in the **outer** position. This minimises the bending forces and therefore reduces the tendency for the angle to flatten out during bending. Use the 'V' tool

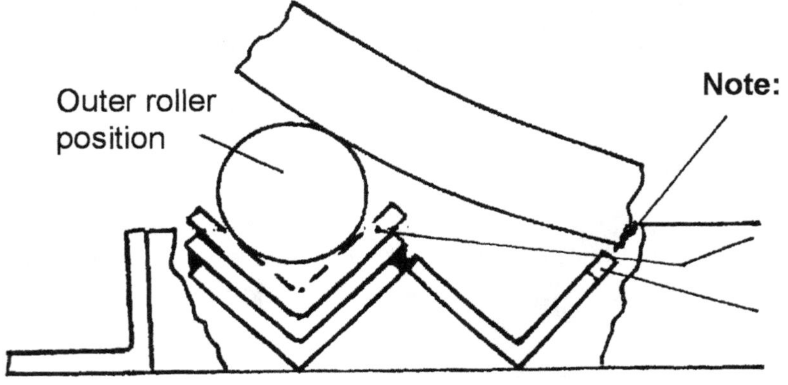


**Note:** rim may make contact with inner rim support. Therefore -

**either:** rest extra piece of angle in support to raise up the roller

**or:** cut a slot out of top edge of angle support.

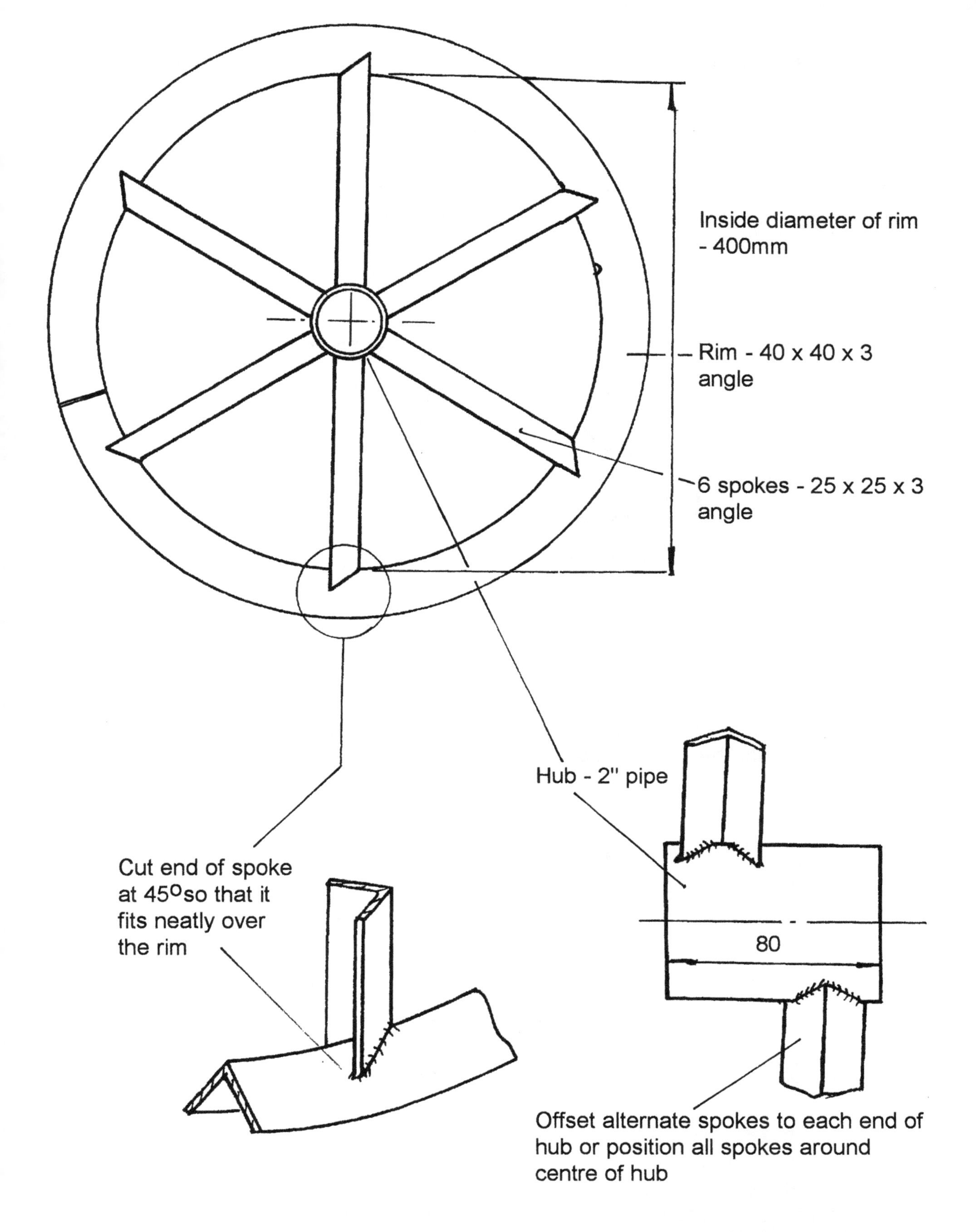


**Figure 5.5: DESIGN OF A WHEEL TO TAKE A 17" MOTORCYCLE TYRE**

- from Chart B, Figure 2.2, the setting of the adjusting screw to give an outside diameter of rim of 456mm is **14 1/2 turns**. Note that at this setting -

  i) 1 turn of screw changes rim diameter by about 45mm;

  ii) for a re-bend, add on 5 turns to the initial setting and add or subtract the change needed to adjust the rim diameter.

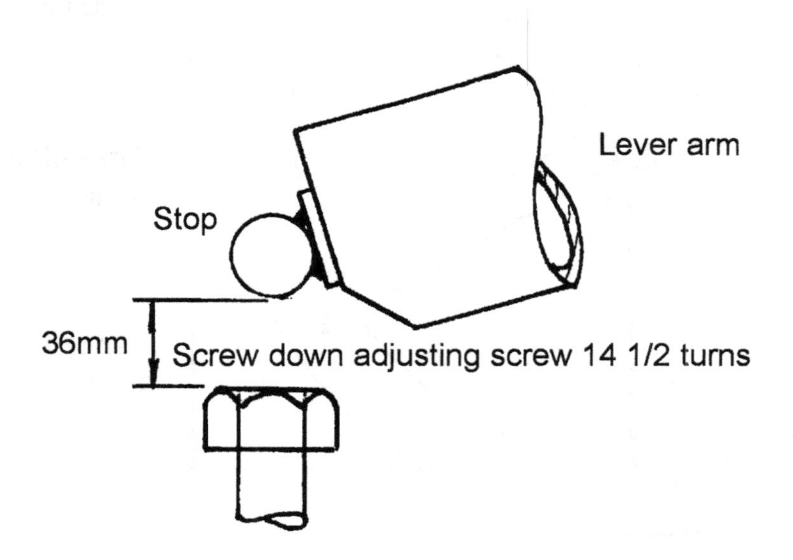


- bend about 250mm at each end of the piece of angle and then remove it from the bender. Cut off the straight pieces at each end;

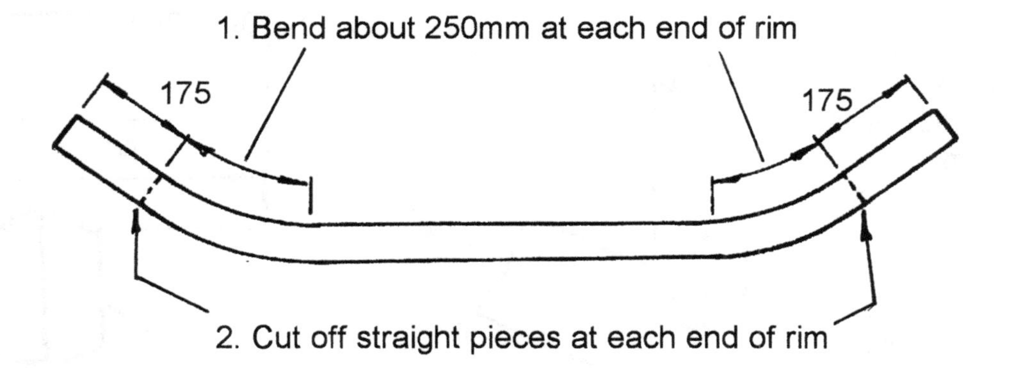


- put the bar back in the bender and complete the bending **make sure** that the rim does not touch the inner roller supports during bending.

**Assembly jig:**

- the inside diameter of the rim is 400mm, therefore the outer surface of the guide post needs to be at a 200mm radius. Use slotted holes in the rim clamp base and use the gauge to position the guide posts - gauge setting:

  = 200 - 10 = 190mm for 20 diameter centre post
  = 200 - 8 = 192mm for 16 diameter centre post

**Hub:**

i) wheel may rotate on a stub axle - for hub design see Figure 6.11;

ii) wheel may be fixed to a live axle - use 1 1/2" medium-wall pipe for the axle.

**Spokes:**

measure and cut 6 spokes from 25 x 25 x 3 angle. Cut one end of spoke at 45° to fit neatly over rim.

**Tyre:**

tyre should be a tight fit on the rim and will need to be levered onto the rim.

**Wheel to take an 18" motorcycle tyre**

The details for making a rim from 40 x 40 x 3 angle to take an 18" motorcycle tyre are as follows:

Inside diameter of rim 424mm

Length of rim = 1405 + 350 = 1755mm

Setting of bender = 13 5/6 turns

# 6. WHEELS FOR MOTOR-VEHICLE TYRES

This Chapter gives details of the design and manufacture of wheels which can be fitted with motor vehicle tyres. The wheel designs cover car tyre sizes from 12" to 16" (Landrover size) and motorcycle tyre sizes 17" and 18". Tyres may be new or used - inner tubes are needed.

**Section 6.1** gives details of two wheel designs to take car and truck tyres. These wheels are designed for loads up to 600kg and are suitable for animal-drawn carts and low-speed trailers.

1. **A split rim wheel** in which the rim is made in two parts, a main part which is connected to the hub or axle and a bolt on part which can be removed to allow the tyre to be fitted or removed.

2. **A detachable bead** wheel in which the rim is in one piece but one bead is bolted on and can be detached to allow the tyre to be fitted or removed.

In both cases the tyre is easily slid onto the rim without the need for it to be levered onto the rim and is held in position by the bolt on part of the wheel. The detachable bead wheel is simpler and quicker to make since only one rim has to be bent and assembled but a wider material section is needed for the rim and more spokes are needed to support the rim.

**Section 6.2** gives details of the construction of axles and hubs and of three axle assemblies which can be used with the above wheels:

- a hub fitted with ball bearings rotating on a fixed stub axle;
- a hub fitted with bush type bearings rotating on a fixed stub axle;
- a live half axle rotating in fixed hardwood block bearings.

The ball bearing assembly is by far the best design and should be selected if the bearings are available and affordable. A similar design can be used for taper-roller bearings.

If ball or roller bearings cannot be used the best alternative is considered to be the live-axle assembly. This uses readily available materials, is simple to construct without the need for a lathe, and is likely to be the lowest cost design. Providing the hardwood block bearings are adequately lubricated and sealed against sand and dust the assembly should have a satisfactory operating life of several years.

**Section 6.3** gives step-by-step details of the construction of the two types of wheels. Many of the steps are the same for the two wheels and the photographs used to show the construction of the detachable bead wheel will also help to illustrate and explain the construction of the split-rim wheel.

**Section 6.4** gives details of the design and construction of a detachable bead wheel to take a motorcycle tyre. The wheel is intended for heavy duty handcarts or light donkey carts.

## 6.1 WHEEL DESIGNS

### 6.1.1 Split-rim wheel (Figure 6.1)

The wheel is made in 2 parts which are clamped together by 4 joining bolts:

PART A - is the main part to which the hub or axle is attached.

PART B - is the "bolt-on" part which is the outer part of the wheel.

The tyre and tube are fitted over PART A and then PART B assembled so that the tyre is held in position. This allows easy assembly and removal of the tyre with the use of only a spanner. It also simplifies puncture repair since only PART B has to be removed and PART A is left in position on the cart.

The design of the wheel may be varied to suit the size of tyre to be used and the materials that are available. Some of the alternatives are as follows:

**Wheel Rim**

The rim is formed from 2 pieces of flat bar - the overall width of the rim should be between 120 and 150mm. Some of the sizes of flat bar which can be used are as follows:

75 wide plus 75 wide
65 wide plus 65 wide
80 wide plus 40 wide
80 wide plus 50 wide

Where unequal widths are used, the wider section should be in PART A of the wheel.

6mm thick material is recommended - for strength considerations the rim section should not be less than 5mm thick - sections greater than 6mm thick will be more difficult to bend and will add unnecessary weight to the wheel.

16mm diameter round bar is preferred for the rim bead but a suitable pipe may also be used.

**Wheel Centre**

The wheel centre joins the rim to the hub or axle. Details of three alternatives are shown as follows:

**Figure 6.1:** Shows the overall construction details of the standard split-rim wheel design which has 4 spokes cut from 50 x 50 x 6 angle.

PART A

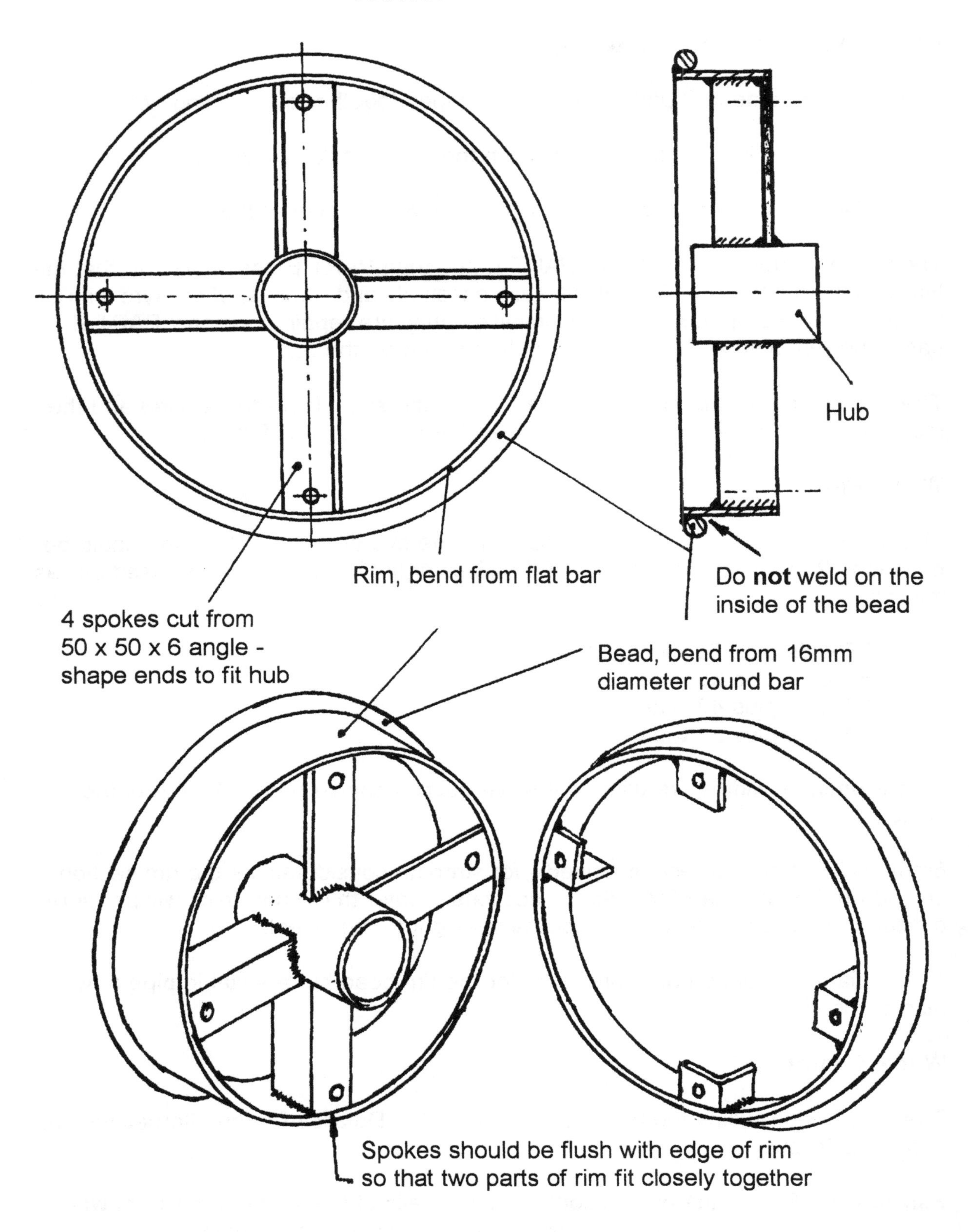


**Figure 6.1: DESIGN OF A SPLIT-RIM WHEEL**

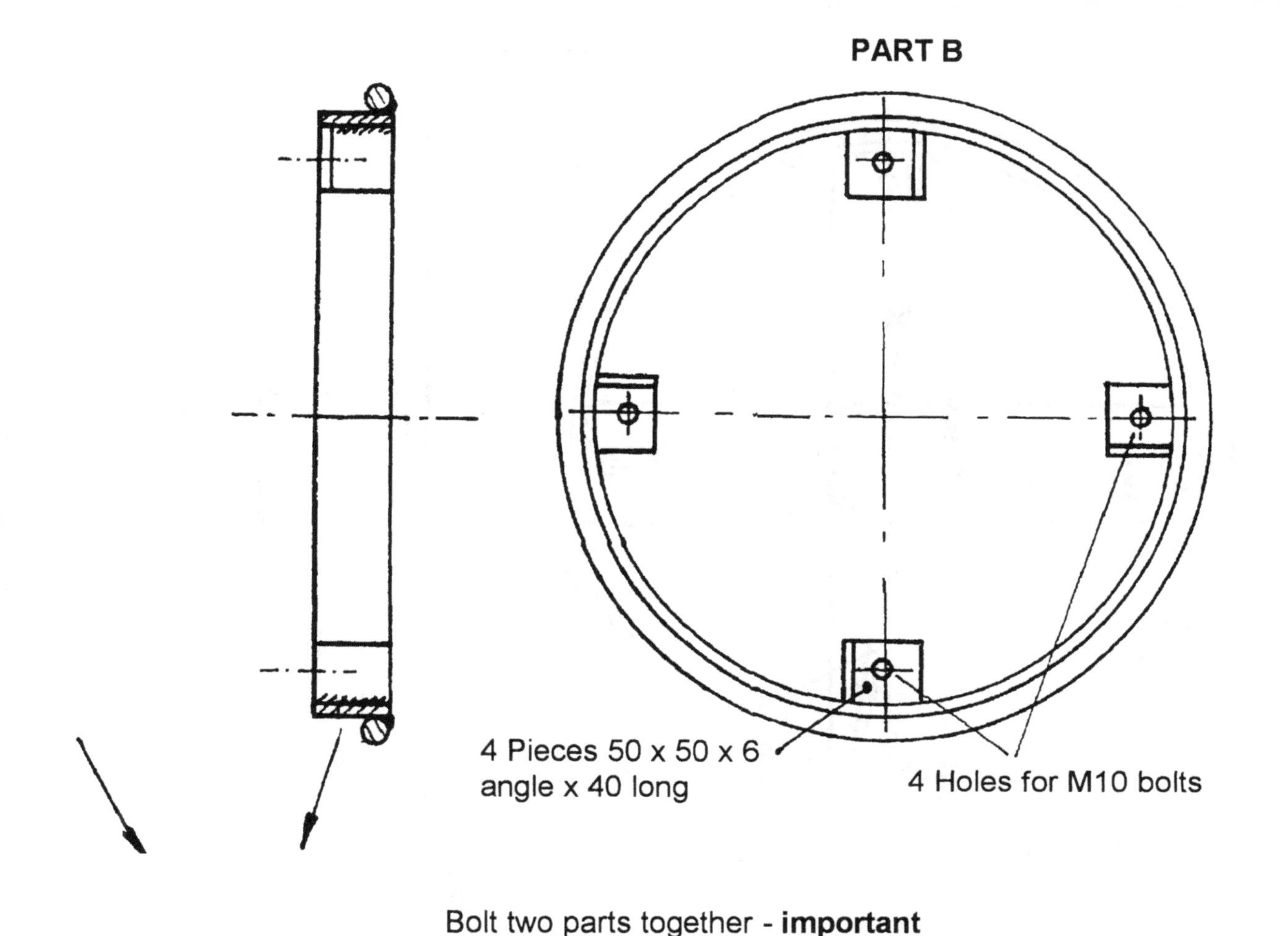


Bolt two parts together - **important** that rims fit closely together so that gap is less than 1mm

Outside diameter of rim about 6mm less than tyre size.

Width of rim 120 to 150mm

**Completed Wheel**

**Figure 6.1: DESIGN OF A SPLIT-RIM WHEEL**

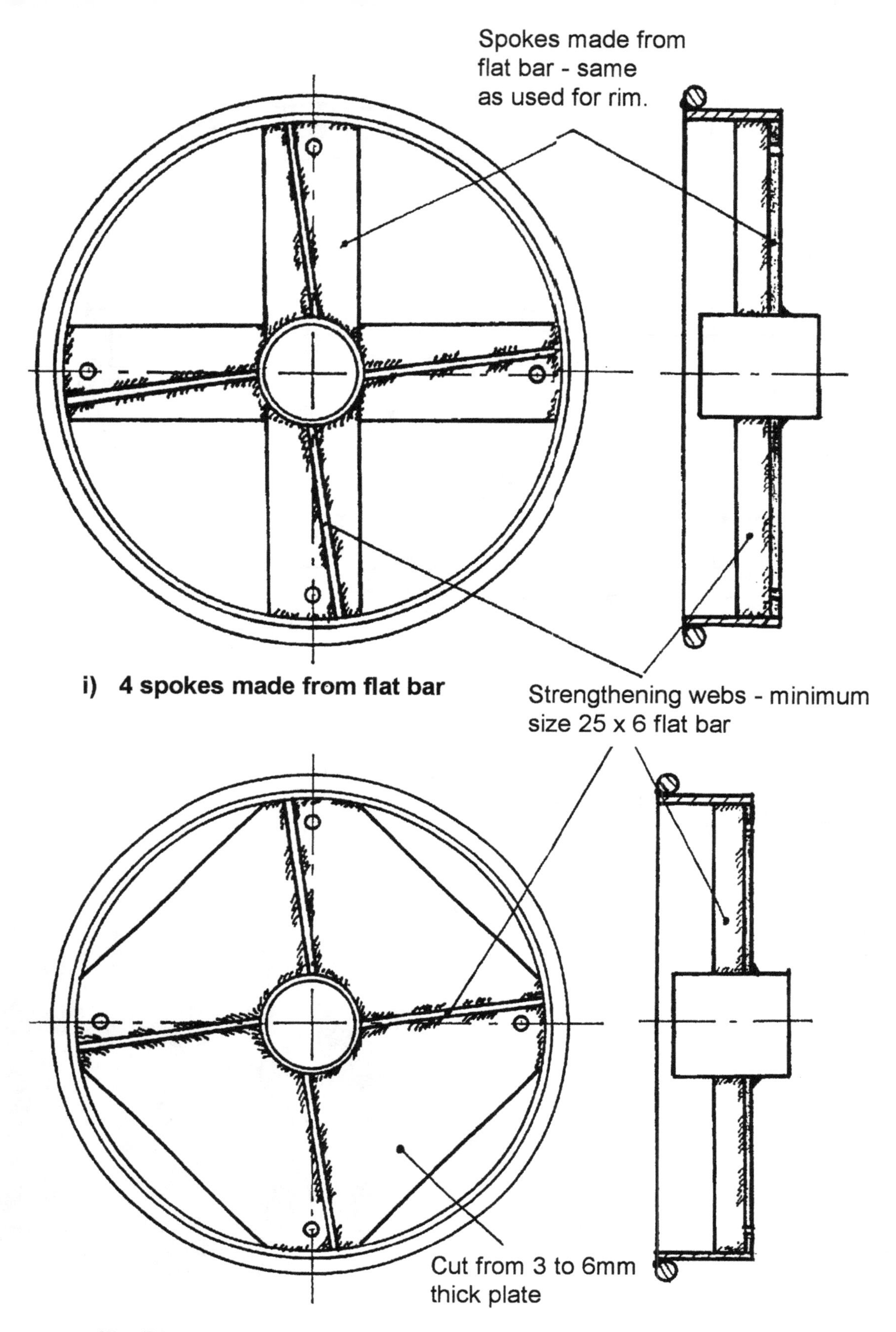


**Figure 6.2: ALTERNATIVE DESIGNS FOR WHEEL CENTRE**

**Figure 6.2:** Shows two other alternatives for the wheel centre.

i) has 4 spokes made from the same flat bar as used for the rim and strengthened with webs of 25 x 6mm flat bar;

ii) has a disc or plate centre cut from 3mm to 6mm thick plate, strengthened with webs cut from 25 x 6mm flat bar.

It should be noted that the alternative designs are for PART A of the wheel. The design of PART B is common to the three alternatives.

### 6.1.2 Detachable-bead wheel (Figure 6.3)

This wheel is simpler to make than the split-rim wheel as only one rim has to be bent and set up in the assembly jig.

The **rim** is formed from 100 x 6 flat bar. If this is not available two lengths of flat bar which give a total width of 100 to 105mm (for example, 2 lengths of 50 x 6 flat bar) may be placed side by side, tack-welded together at their ends and then formed together in the rim bender. When the rim has been bent the two pieces of bar should be more fully welded together on the **inside** surface of the rim, but leave gaps in the weld where the rim will fit over the guide-posts in the assembly jig. **The rim width should not be less than 100mm**

The preferred design of the wheel centre is to use 6 spokes of 50 x 50 x 6 angle section. More spokes are needed to support the rim than for the split-rim wheel because the rim is only reinforced on one edge by a welded on bead. A plate or disc centre (see Figure 6.2) may also be used but 40 x 6 or 50 x 6 flat bar should be used for the webs.

The **beads** may be 16 or 20 diameter round bar or a similar size tube. The beads should be set right at the edge of the rim, half-on and half-off the rim, to give the greatest width between the beads to fit the tyre. The bead should only be welded to the rim on the outside but strong welds are needed. **Strong welds** are also very important between the flat pads and the bead on the detachable bead.

The same hub and axle assemblies may be used for both the split-rim and detachable-bead wheels. These are detailed in Section 6.2.

## 6.2 HUB/AXLE ASSEMBLY

### 6.2.1 Construction of hub and axle

The design of the hub/axle assembly depends mainly on what type of bearing is available or affordable and whether a lathe is available to machine the hub and axle. A general discussion of the options and selection of the most appropriate designs has been given in Chapter 4. **Figures 6.4(i) to (iv)** show some ideas for making the hub and axle. These are for stub axles where short lengths of axle are welded into each end of an axle beam which is attached to the frame of the cart. The designs shown are:

i) **If a lathe is available** - the axle can be machined from "black" mild steel round bar. The hub can be machined from thick-wall pipe (steam pipe) or a cylinder fabricated from flat bar using blacksmithing methods;

ii) **If a lathe is not available** - the axle can be made from bright mild steel round bar (sometimes known as shafting). The surface, roundness and dimensional accuracy of this bar is usually good enough for it to be used without machining. A length of threaded rod, M16 or M20, can be welded centrally to one end of the axle for the nut to hold the wheel in place. The hub can be made from medium wall steel pipe (water pipe) by slitting it along its length, squeezing it up (or opening it out) so that the bearings are a tight fit in it and then welding along the slit.

**Axle-beam**: suitable axle-beams for various vehicle loads are listed below:

| All-up Vehicle Weight | Suitable Design of Axle-Beam |
|---|---|
| 400-500kg<br>(single donkey cart or<br>2 man hand cart) | • 40 x 40 x 6 angle section<br>• Box formed from 2 lengths of 40 x 40 x 4 angle. |
| 700-800kg<br>(cart for pair of donkeys) | • 50 x 50 x 6 angle section<br>• Box formed from 2 lengths of 40 x 40 x 6 angle. |
| 1,000-1,200kg<br>(ox-cart) | • 60 x 60 x 6 angle section<br>• Box formed from 2 lengths of 50 x 50 x 6 angle. |

## 6.2 HUB/AXLE ASSEMBLY

### 6.2.1 Construction of hub and axle

The design of the hub/axle assembly depends mainly on what type of bearing is available or affordable and whether a lathe is available to machine the hub and axle. A general discussion of the options and selection of the most appropriate designs has been given in Chapter 4. **Figures 6.4(i) to (iv)** show some ideas for making the hub and axle. These are for stub axles where short lengths of axle are welded into each end of an axle beam which is attached to the frame of the cart. The designs shown are:

i) **If a lathe is available** - the axle can be machined from `black´ mild steel round bar. The hub can be machined from thick-wall pipe (steam pipe) or a cylinder fabricated from flat bar using blacksmithing methods

ii) **If a lathe is not available** - the axle can be made from `bright´ mild steel round bar (sometimes known as shafting). The surface, roundness and dimensional accuracy of this bar is usually good enough for it to be used without machining. A length of threaded rod, M16 or M20, can be welded centrally to one end of the axle for the nut to hold the wheel in place. The hub can be made from medium wall steel pipe (water pipe) by slitting it along its length, squeezing it up (or opening it out) so that the bearings are a tight fit in it and then welding along the slit.

**Axle-beam**: suitable axle-beams for various vehicle loads are listed below:

| All-up vehicle weight | Suitable design of axle-beam |
|---|---|
| 400-500kg<br>(single donkey cart or<br>2-man handcart) | • 40 x 40 x 6 angle section<br>• Box formed from 2 lengths of 40 x 40 x 4 angle |
| 700-800kg<br>(cart for pair of donkeys) | • 50 x 50 x 6 angle section<br>• Box formed from 2 lengths of 40 x 40 x 6 angle |
| 1000-1200kg<br>(ox-cart) | • 60 x 60 x 6 angle section<br>• Box formed from 2 lengths of 50 x 50 x 6 angle |

1. **Hub made from thick wall (steam) pipe**

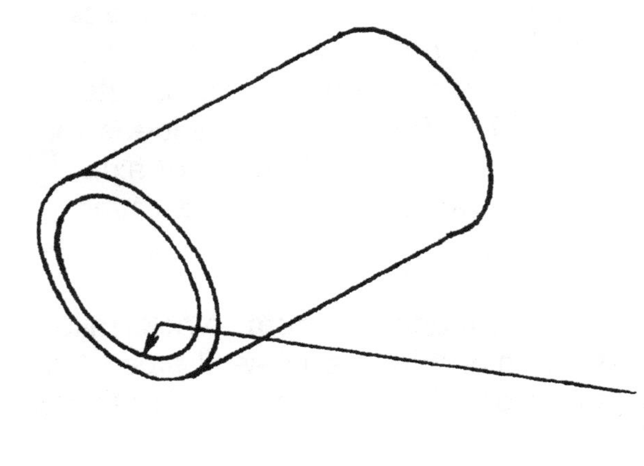


2. **Hub fabricated from flat bar**

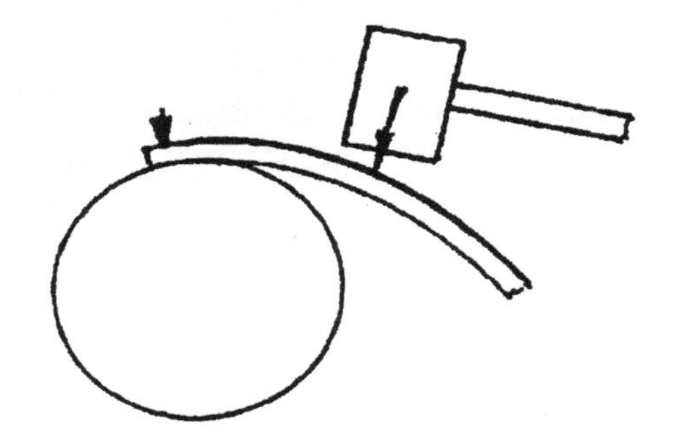


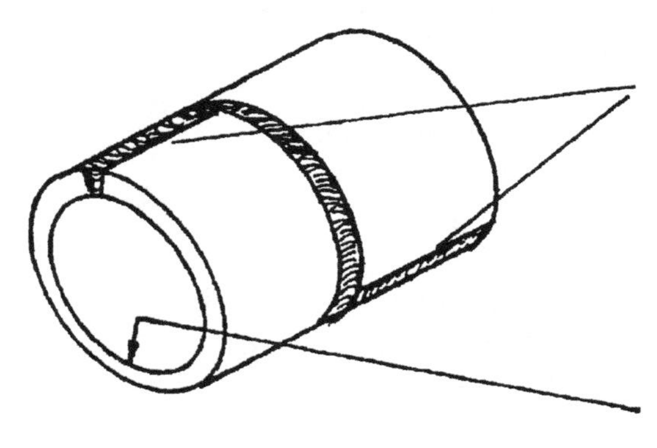


**Figure 6.4(i): MANUFACTURE OF HUBS**

## 2. Hub made from medium gauge water pipe

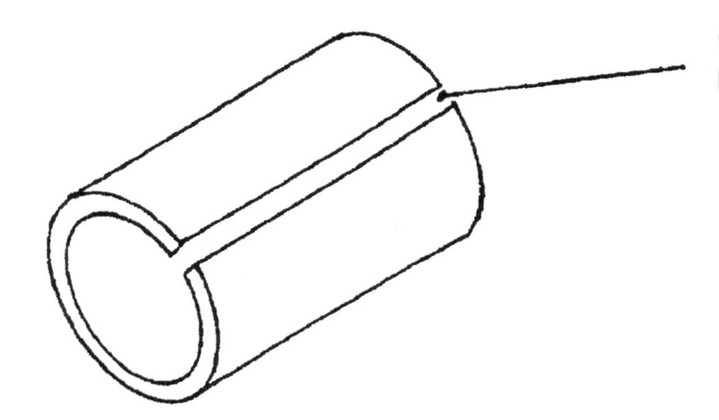

Cut slot along length of hub removing the seam of the pipe.

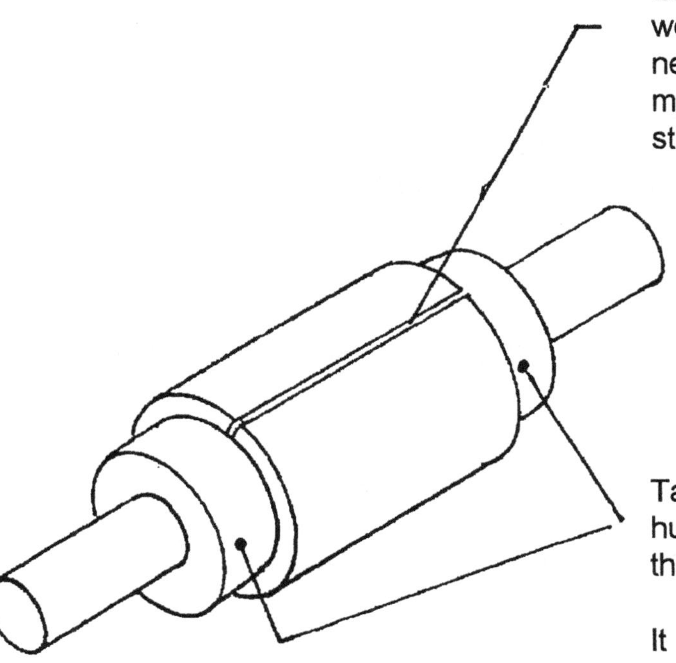

Squeeze up the slot so that the bearings are a tight fit in the hub. Leave a gap of about 1 to 2mm for welding. (In some cases it may be necessary to open up the slot to make the hub larger and to weld a strip into the slot).

Tap the bearings into each end of the hub using a length of axle to line up the bearings.

It is a good idea to use a pair of scrap bearings or dummy bearings so that the bearings may be left in position during welding. (A dummy bearing is a piece of bar machined to the same size as the bearing).

Weld up the slot - if good bearings are being used, tack-weld the slot and remove the bearings before completing the weld.

**Figure 6.4(ii): MANUFACTURE OF HUBS**

1. **Stub axle machined from a length of round bar**

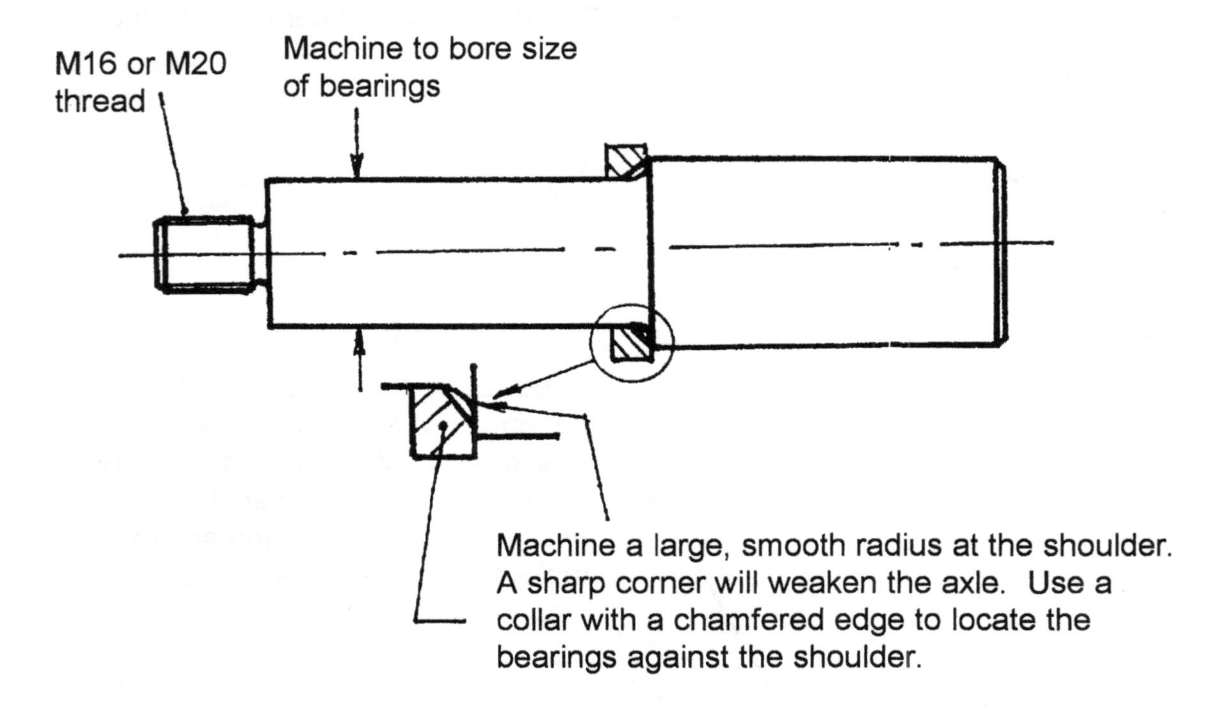


2. **Stub axle cut from Bright Mild Steel Bar (BMS) - no machining needed**
(sometimes called "shafting")

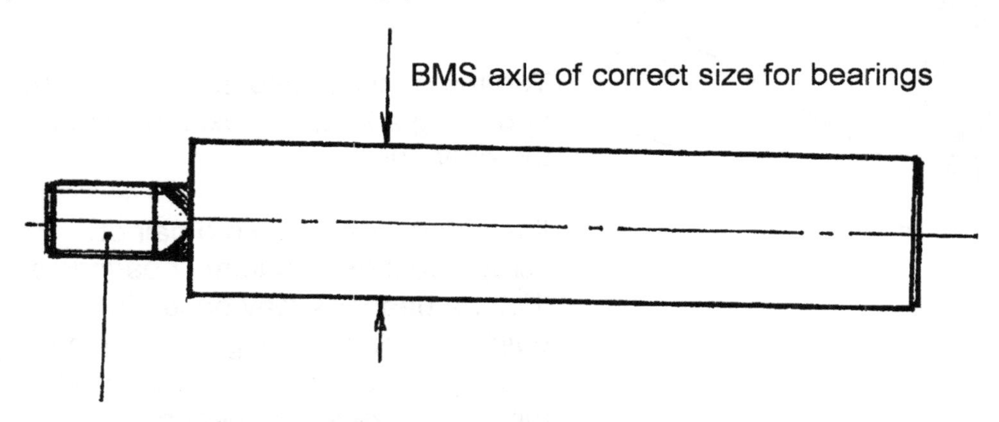


**Figure 6.4(iii): MANUFACTURE OF AXLES**

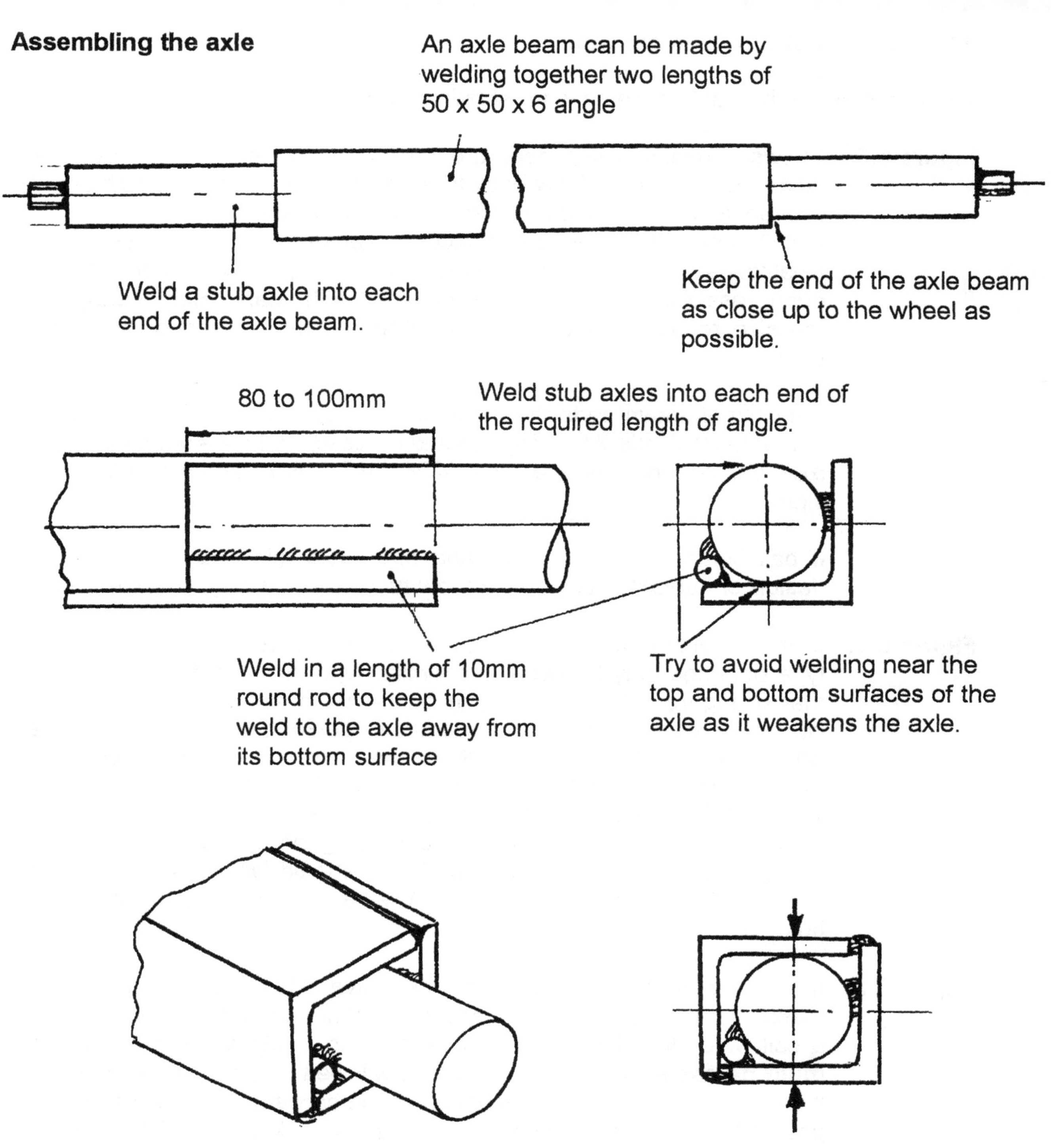


**Note:** an alternative axle-beam can be made from a single length of 60 x 60 x 6 angle.

**Figure 6.4(iv): MANUFACTURE OF AXLES**

### 6.2.2 Design of hub-axle assembly

Three designs of hub-axle assembly are shown in the following figures:

**Figure 6.5:** Shows a design using deep-groove ball bearings. These have the advantages of being fairly easy to assemble and they can be obtained with seals and lubricated for life. Taper-roller bearings are usually used on motor-vehicle axles because they can support larger axial loads. However, they are more costly, require careful adjustment during assembly and cannot be obtained with seals. Ball bearings of the size recommended are quite adequate for low-speed carts and trailers.

For animal-drawn carts the minimum axle diameter should be 40mm and at least Grade 2 bearings should be used i.e. for a 40mm diameter axle the bearings will be type 6208 2RS (2RS means sealed both sides).

If ball bearings are available and affordable this is by far the best design of hub/axle assembly for low friction and a long, reliable life.

**Figure 6.6:** If ball or roller bearings are not available or not affordable then bush-type bearings may be used in a similar type of hub/axle assembly to that of Figure 6.5. However, in bush-type bearings there is sliding contact between the bush and axle and friction and wear in the bearings will be much higher than for ball bearings. Wear is the major problem for this type of bearing and over a period of time the "wobble" of the wheel on the axle will gradually increase until it becomes unacceptable. However, if the bearings are properly designed and constructed and **adequately lubricated and sealed** they should last for several years (at least 10,000km of operation). A lathe is needed to machine the bushes.

**Figure 6.7:** If a lathe is not available an alternative which does not require machining is to make a bush by winding 6mm diameter round rod into a coil which fits closely over the axle. The inside surface of the coil **must be case-hardened** otherwise it will wear rapidly. The coil bush performs very well because it holds grease between the coils and gives good lubrication of the bearing surfaces over a long period of time.

**Figure 6.8:** This shows an assembly in which a live axle rotates in two fixed block bearings made from hardwood. The advantage of this arrangement is that because the bearings are spaced far apart the effect of wear on the "wobble" of the wheel is much lower than for a hub assembly - this type of assembly should last for up to 10 times longer than a hub assembly using hardwood bushes. The disadvantage is that the bearings are more difficult to line up. For this reason it is better to have a slightly larger initial clearance in the bearing to allow for small misalignments and shrinkage of the wood - about 2mm is recommended (on the diameter). Lining up the bearings is helped by fitting them in a channel formed from two lengths of angle. The hardwood blocks should be fitted tightly into the channel to reduce any tendency for the hardwood to split.

The advantages of hardwoods are that they are usually readily available at reasonably low cost and that they can easily be machined.

**Figure 6.5: HUB ASSEMBLY USING TWO BALL BEARINGS**

The figure shows a hub assembly for a 40mm diameter stub axle and two Type 6208 ball bearings. The axle is a 190mm length of 40mm diameter Bright Mild Steel (BMS) bar with a M16 or M20 stud welded to one end.

It is best to use **ball bearings** which are sealed and lubricated for life, if these are available. These are specified as Type 6208 2RS (2RS means sealed on both sides). Scrap bearings may be used to reduce cost but these should rotate freely without excessive sticking or tight spots. If new bearings are used it may be possible to obtain lower cost types imported from China, India, Korea or Eastern European countries.

The **hub** is made from 3" medium-wall water pipe. Cut a slot along the length of the pipe about 4 or 5mm wide, removing the pipe seam. Squeeze up the pipe so that the bearings can be tapped tightly into the ends. A gap of 1 to 2mm should be left so that a good penetration weld is obtained.

The **outer sleeve** is to locate the bearings in the hub. Drill a large hole (at least 13mm diameter) near the centre of the hub, fit the sleeve **centrally** inside the hub and plug-weld it in position.

The **bearing spacing sleeve** is to prevent the bearings being damaged when the end nut on the axle is tightened. It should be 1 or 2mm longer than the outer sleeve.

It is important that the ends of the hub and both sleeves are made square with the axis of the pipe otherwise the bearings and/or the hub may be pulled out of line during assembly.

It is **very important** to fit dust caps over each end of the hub to keep out sand and dirt. This will greatly increase the life of the bearings. The caps should be as close fitting as possible.

If **unsealed bearings** are used it will be necessary to provide a means of lubricating the bearings. Drill a 11mm diameter hole at a convenient position through the hub and sleeve and weld an M10 nut over the hole. Add grease through the hole and plug the hole with an M10 bolt.

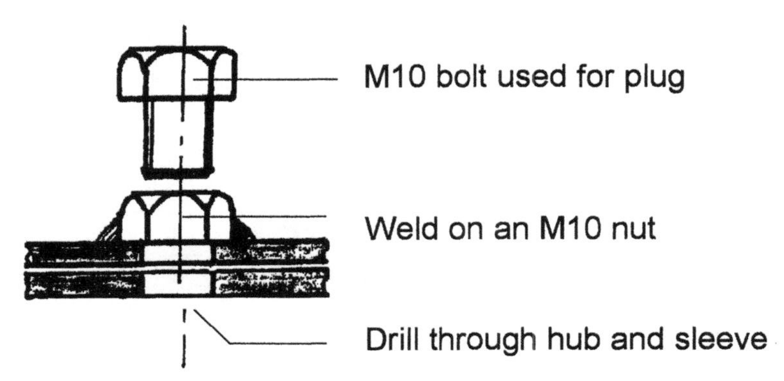

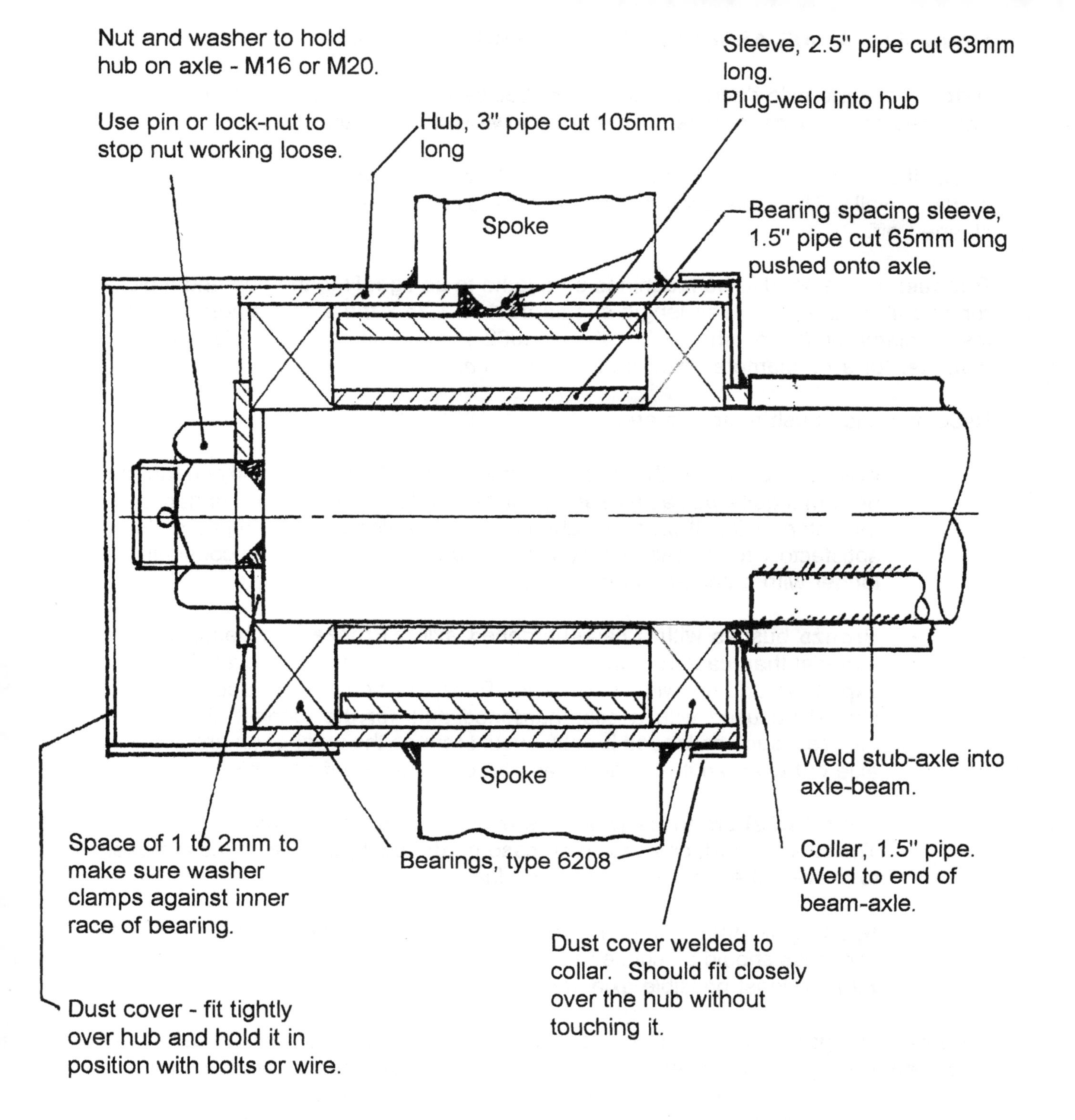


**FIGURE 6.5: HUB ASSEMBLY USING TWO BALL BEARINGS**

**Figure 6.6: A HUB ASSEMBLY WITH BUSH-TYPE BEARINGS**

**Axle:** the same axle design is used as for ball bearings with stub axles welded into each end of a box-beam axle. The size shown is 40mm diameter .

**Hub:** if pipe is used, the preferred size is 2.5" medium wall - second choice is 2" medium wall. The pipe does not need to be slit but the inside seam should be cleaned off.

**Bushes:** these need to be machined on a lathe to be tight fit in the hub and a close running fit on the axle. The length should be 1 to 1.25 x the axle diameter and the inside diameter (bore) should be about 1.002 x axle diameter i.e. clearance on diameter for a 40 diameter axle of 0.05 to 0.1mm.

Recommended bush materials are:

- **cast-iron** bushes with a case-hardened axle. Cast-iron is fairly cheap and performs quite well as a bearing but it is hard and will wear an unhardened mild steel axle. If properly lubricated and sealed the bearings should be satisfactory for at least 5000km with an unhardened axle and considerably longer with a case-hardened axle;

- **Bronze** bushes with a case-hardened axle. Bronze is a better bearing material than cast-iron and should give a longer life. However, it is quite expensive and not widely available. Phosphor bronzes are quite hard and should be used with a case-hardened axle for the longest life. Softer bronzes, such as leaded bronze, would reduce the wear of an unhardened axle but even so it is best to case harden the axle if possible;

- **Thrust washers** these should be machined from 5 or 6mm thick flat bar. The rubbing surfaces should be case hardened if possible. Pins are used to prevent the washers rotating on the axle;

- **Spacing washer** is fitted between the inner thrust washer and shield. Its thickness should be chosen so that the hub is free to rotate on the axle but with the least possible axial movement.

**Lubrication:** good lubrication of metal bushes is **essential** otherwise bushes will wear very rapidly and may seize up. Grease is the best lubricant because it stays in place. An inlet for grease is provided in the hub mid-way between two spokes. The space inside the hub between the bushes should be filled with grease. Shallow grooves are provided in the bush and on the thrust washer to spread the grease over the bearing areas.

**Sealing:** proper sealing of the hub is very important to keep out sand and dirt. The end cap must fit tightly over the outside of the hub - it can be held in position by bolts or by wiring it to the spokes. The inner shield should have a small clearance, about 2mm, over the hub which may be sealed with grease and a strip of scrap inner tube wound around the hub (see Figure 4.3).

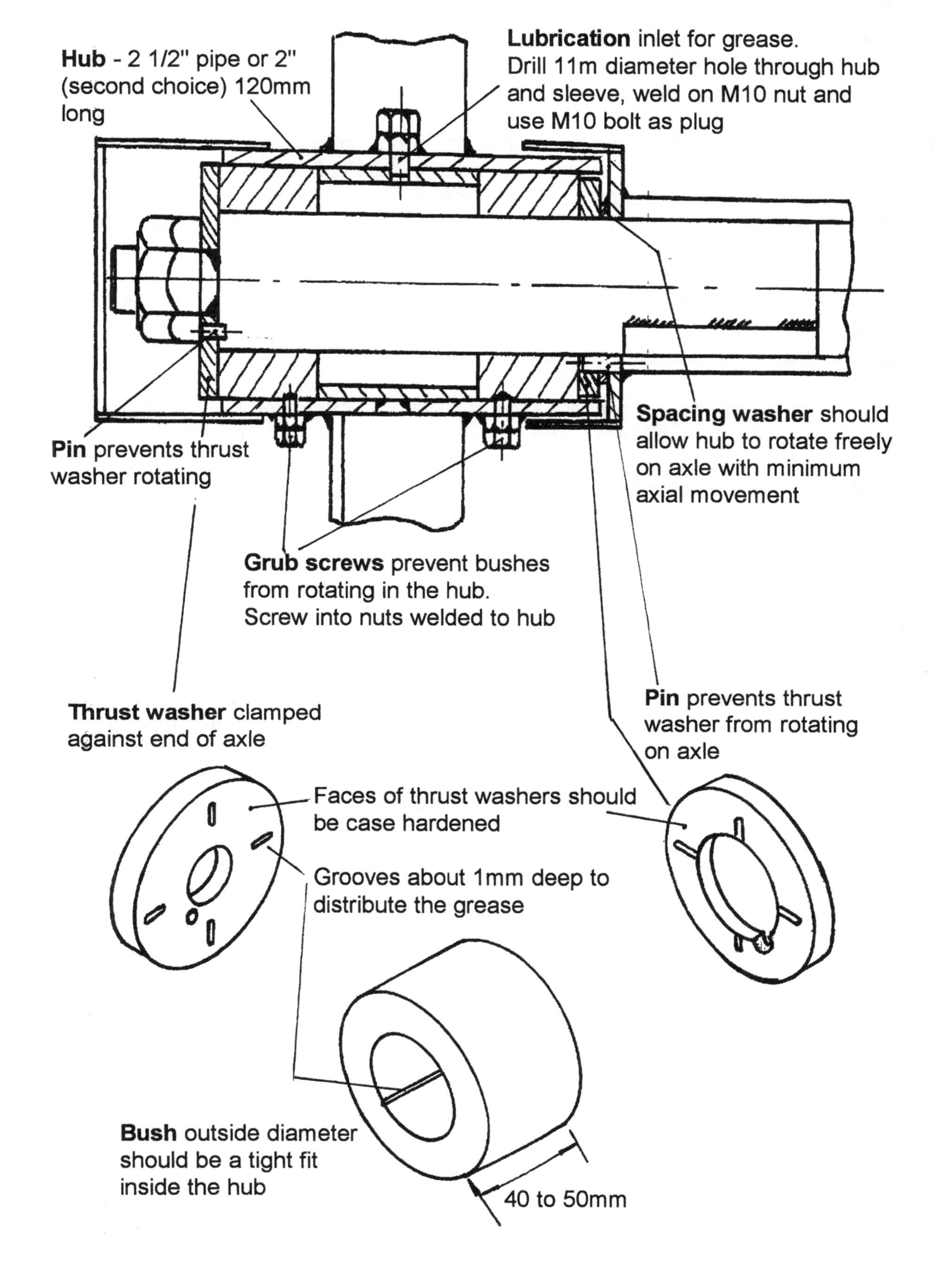


**Figure 6.6: DESIGN OF A HUB ASSEMBLY FOR BUSH BEARINGS**

**Figure 6.7: HUB ASSEMBLY WITH COIL BUSHES**

**Axle:** same as for previous design - 40mm diameter.

**Hub:** cut from 2" medium-wall pipe. Cut a slot about 3mm wide to remove the seam of the pipe, squeeze up the pipe so that the coil bushes are a tight fit inside the hub and then weld up the slot. A sheet metal ring is welded over the inner end of the hub to prevent sand or dust falling directly into the thrust bearing.

**Coil bushes:** these are wound from 6mm diameter round bar. They should be wound onto a former which is a slightly smaller diameter than the axle so that when the coil "springs back" it is a good running fit on the axle. **The inside surface of the coil must be case-hardened**.

**Thrust washers:** these are similar to the previous design. They are prevented from rotating on the axle by pins. The outer pin is fitted into a hole in the end of the axle and the inner pin is welded into a hole in the dust cover. The thrust washers rub against plain washers tack welded to the ends of the coil bushes. The rubbing faces should be case hardened if possible.

**Spacing washer:** this is fitted between the inner thrust washer and shield. Its thickness should be chosen so that the hub rotates freely on the axle with the least possible axial movement.

**Lubrication:** the inside of the coil bushes should be packed with grease and the cavity between the bushes also filled with grease. The grease should stay in position for a long period of time but a filler hole is provided to add additional grease as needed.

**Sealing:** this is the same as for the previous design.

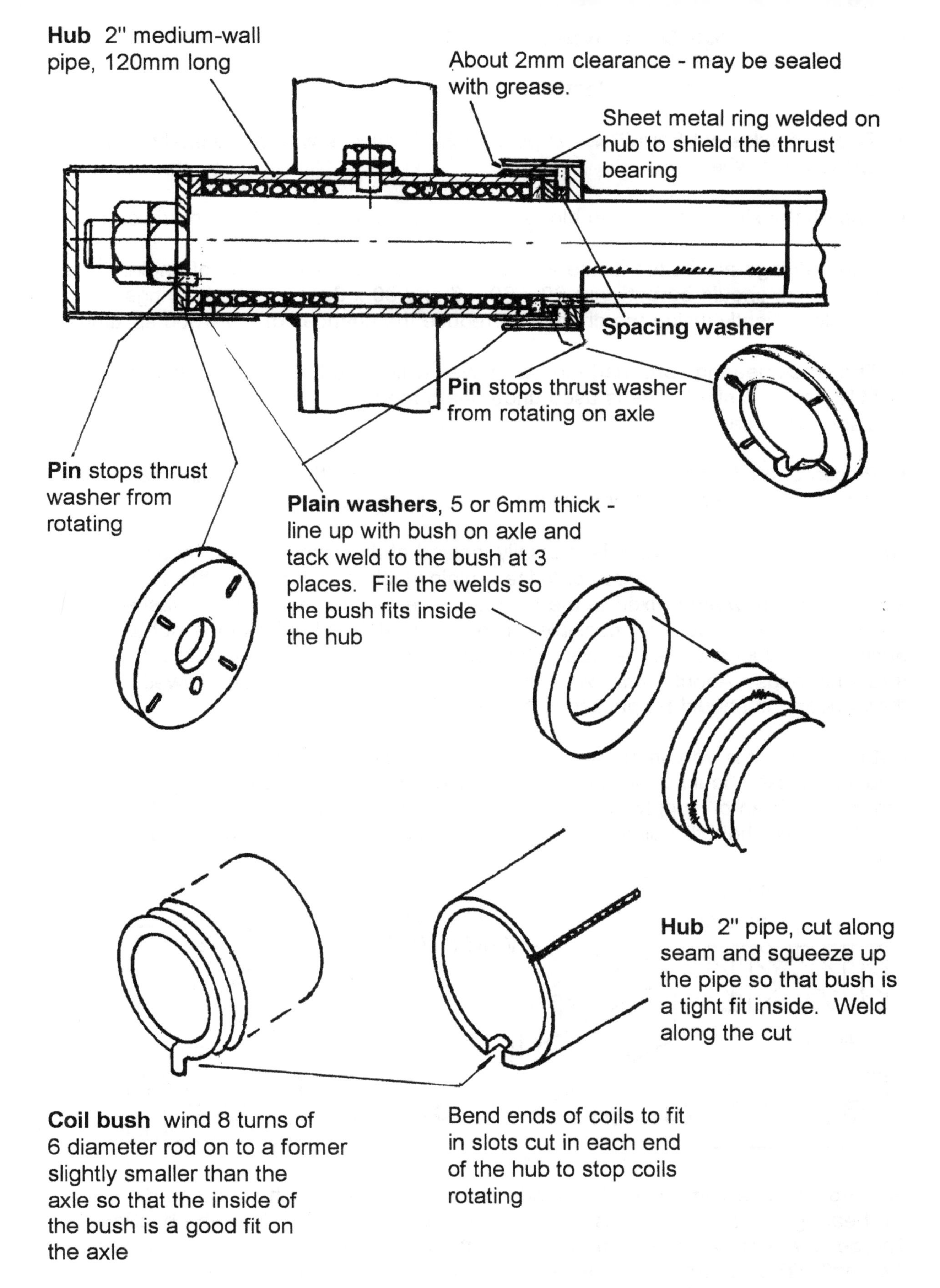


**Figure 6.7: HUB ASSEMBLY WITH COIL BUSHES**

**Figure 6.8: DESIGN OF A LIVE-AXLE ASSEMBLY**

The design has the following features:

- Two half axles cut from 2" water pipe each of which is welded centrally to the spokes of a wheel
- Each half-axle rotates in two block-bearings made from a suitable hardwood
- The bearings are bolted inside a channel made from two lengths of angle. The angle size should be between 60 x 60 x 6 and 80 x 80 x 6. The bearings should fit tightly inside the channel to reduce the chances of the wood splitting
- The block-bearings are made in two halves which are bolted together with two M16 bolts. A steel plate is used underneath the bearing to evenly distribute the clamping force
- Pieces of angle are welded to the channel beam on either side of the bearings to protect the bearings from striking rocks etc. on rough ground.

**Axial location:** each half-axle is axially located at the inner bearing by thrust washers. The thrust washers comprise a plain washer rotating with the axle against a spiral washer fixed to the bearing. The contact faces of the washers should be case hardened if possible. The end washer is held on by a nut to allow accurate axial location of the axle. Care should be taken not to clamp up the thrust bearings. About 1mm axial movement of the axle should be allowed. The thrust bearings should be packed with grease.

**Lubrication:** it is very important to keep the bearings well lubricated. The wooden blocks should be initially boiled and soaked in oil. Each bearing is provided with an oil supply, and oil reservoirs filled with material which will soak up oil. These will hold the oil for a long time but periodically new oil will have to be added.

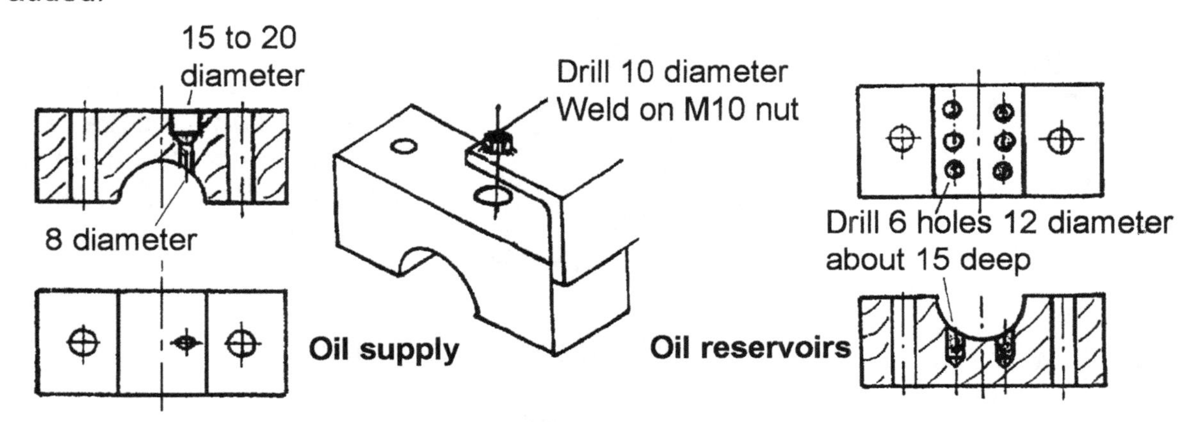


**Sealing:** it is very important to provide seals to keep as much sand and dirt out of the bearings as possible. This will considerably lengthen the life of the bearings. The seals are of the shield type made up of a dust cover fitted over a disc fixed to the shaft. The covers should be screwed tightly against the face of the bearing.

**Note:** at the outer bearing the disc seals are clamped to the axle to avoid welding which would significantly weaken the axle.

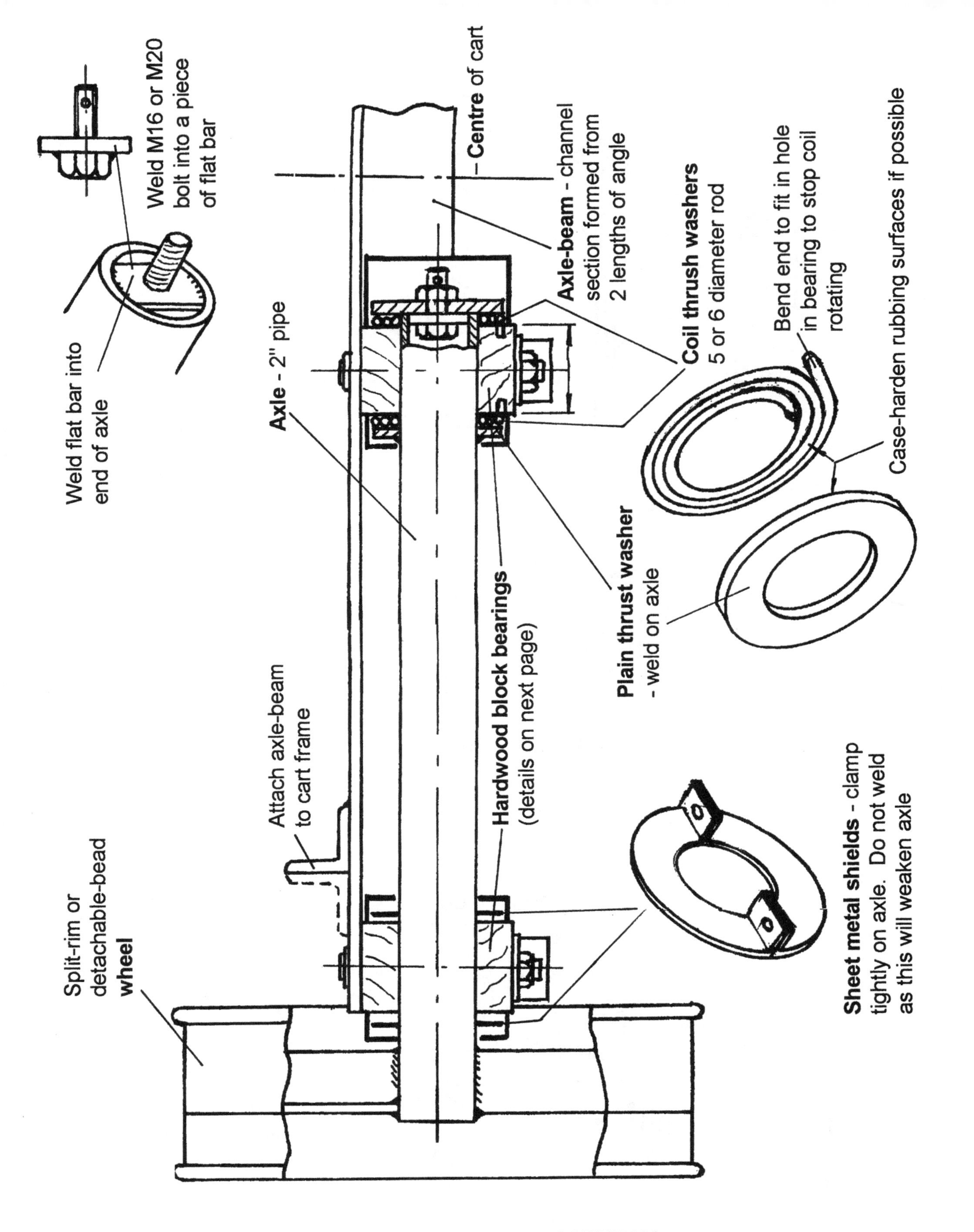


**Figure 6.8(i): LIVE-AXLE ASSEMBLY**

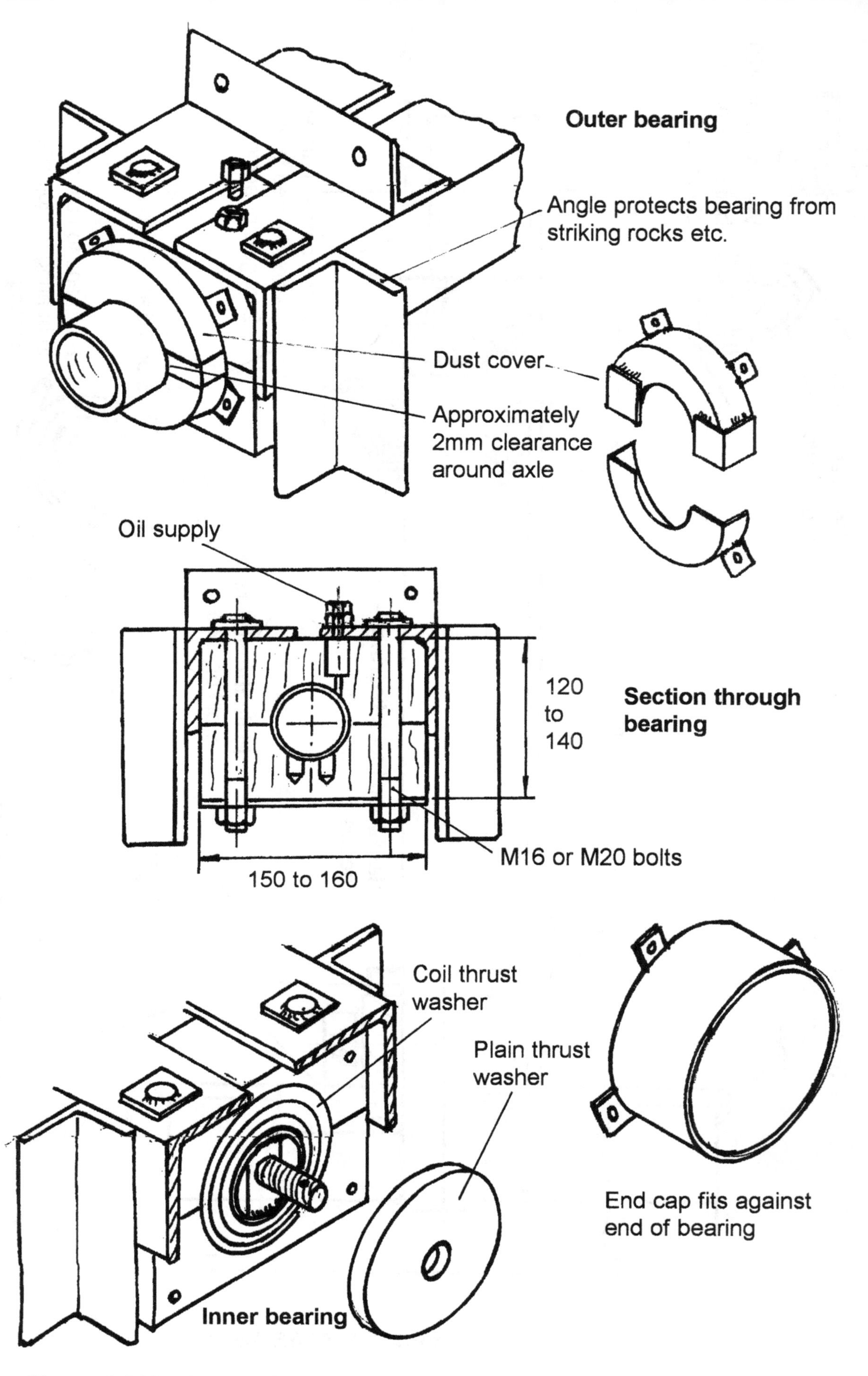


**Figure 6.8(ii): DETAILS OF BEARINGS FOR LIVE-AXLE ASSEMBLY**

## 6.3 CONSTRUCTION OF WHEELS

This section gives step-by-step instructions for construction of a split-rim wheel and detachable bead wheel. The instructions given in **6.3.1** are illustrated by the photographs in **6.3.2**. General instructions on the use of the equipment are also given in **Section 2.3**

### 6.3.1 Split-rim wheel

**1. Prepare the material for the rim**

**1.1** Cut lengths of flat bar for the rim and round bar for the beads. The lengths for different tyre sizes are as follows:

| Tyre size | Outside diameter of rim | Length of bar for rim | Length of bar for bead |
|---|---|---|---|
| 13" | 324mm | 998mm | 1290mm |
| 14" | 350mm | 1078mm | 1370mm |
| 15" | 375mm | 1158mm | 1450mm |
| 16" | 402mm | 1242mm | 1530mm |

**Notes:**

- The **rim** lengths are the estimated lengths needed with a 2mm gap at the joint. For the first trial it is best to cut the bar 10 to 20mm longer and to determine the exact length needed from the length trimmed off
- The lengths of the **beads** have an allowance of 110mm at each end which will not be formed during the bending operation and will need to be cut off after bending.

**1.2** Mark out the length of the flat bar in steps of 25mm (1") with chalk marks - see photograph. Mark out the bars for the beads in a similar manner.

**2.** **Bending the rim** (see also Section 2.3 and photographs in 6.3.2).

**2.1** Set the rollers in the **inner** position.

**2.2** Fit the 2-point bending tool in the machine.

**2.3** Zero the machine - to do this put a **straight** piece of the rim material in the machine and pressing **lightly** down on the lever socket screw the stop upwards till it just touches the lever arm.

**2.4** Screw down the stop the number of turns needed for the size of rim to be made. The settings obtained from the graph in **Figure 2.2A** are as follows:

| Tyre size | Number of turns for rim | Number of turns for bead |
|---|---|---|
| 13" | 6 1/4 | 5 3/4 |
| 14" | 5 5/6 | 5 1/3 |
| 15" | 5 1/2 | 5 |
| 16" | 5 | 4 2/3 |

**Note:** these settings are a guide and the exact number of turns needed will depend on the quality of steel used.

**2.5** Use the above settings for the initial trial. If the rim looks as if it is coming out the wrong size do not stop, complete the bending and measure the outside diameter of the rim. If it is within about 5mm of that needed the rim can be used. If it is bigger or smaller than this then change the number of turns of the stop - as a guide, **1/6** of a turn will change the rim diameter by about **12mm**, and **1** turn by about **72mm**.

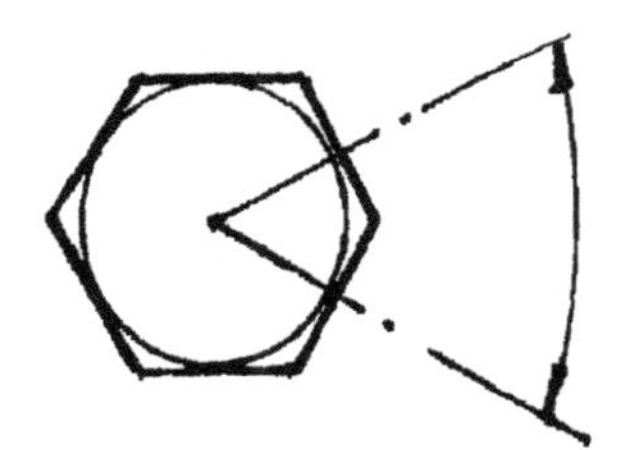

1/6 turn changes rim diameter by about 12mm

**2.6** Bend further lengths of bar until you get the correct size - 2 or 3 trials should be enough to get the size right. Once you have the correct setting it is best to bend up a batch of rims before changing the setting. The trial rims of the wrong size can be corrected.

Rims that are too large can be passed through the bender a second time **but** the settings of the stop will have to be taken from the **re-bends** curve in Figure 2.2A. This means that the number of turns will need to be increased by 1 5/16 **plus** the number of turns needed to reduce the diameter. For example if you are making a rim for a 16" tyre and the rim comes out at 420mm diameter then to correct it you will need to increase the number of turns by $1\ 5/6 + \frac{(420-400)}{72} = 1\ 5/6 + 3/12 = 2\ 1/12$.

**Note:** an alternative method is to use the chart as explained in Section 2.3.

If the trial rim is too small then hammer all around the rim to open it up so that it can be re-bent in the bending machine.

**2.7** When the rim has been bent to the correct size, remove it from the bender and hammer the two straight ends over an anvil or former to complete a smooth circle.

**2.8** When operating the bender, start with the leading edge of the rim just resting on the furthest roller and pass the bar through the machine in the steps shown by the chalk marks. At each step push the lever arm down until it just touches the stop.

- **do not miss any marks**

- **make sure the lever arm touches the stop but do not press down too hard on the stop**

- **keep the bar accurately lined up so that it is bent properly and the two ends line up.**

3. **Bending the Beads**

3.1 It is best to bend round bar with the 'V'-tool, therefore fit this tool in the machine. Leave the rollers in the inner position.

3.2 Zero the machine using a straight length of the bead material. Then screw down the stop by the number of turns specified in **2.4** for the tyre size.

3.3 Bend up the beads in a similar manner to that used for the rims.

4. **Assembling the Wheel** (see Figures 6.9 and 6.10).

4.1 Set up the assembly jig by bolting the rim guides in the correct positions for the size of wheel. The positions of the guide stops are shown in **Figure 6.9.**

4.2 Take the rim for PART A of the wheel - if the two parts of the rim are of different widths PART A will be the **wider** rim.

4.3 Fit the rim over the guide posts, locating the joint midway between guides 2 and 3. Clamp the rim against the posts, tightening the clamps in the order shown in **Figure 6.9.** Make sure the rim sits evenly on the base of each rim guide - if it does not, clamp the rim down onto the base pieces with the two clamping bars.

4.4 When the rim is clamped in position there should be a gap of about 2mm between the ends for a good penetration weld. If the gap is too small remove the rim and cut a small piece off one end of the rim. If the gap is too large fit a piece of bar in it so that a strong weld can be made at the joint. Weld up the joint.

4.5 Make up the wheel hub and fit scrap or dummy bearings. Fit the hub over the centre-post of the jig as shown in **Figure 6.9** using the collars to make sure it is correctly positioned in the centre of the rim. The hub must also be at the correct height relative to the rim.

**Note:** machined collars or bosses may be fitted into each end of the hub to locate it on the centre post instead of scrap or dummy bearings.

4.6 Measure the length of spokes needed and cut **4** lengths of **50 x 50 x 6** angle for the spokes (or other sections for alternative designs). Shape the inner ends of the spokes so that they fit neatly against the hub for welding. Drill holes for the clamping bolts.

## View on top of the assembly jig

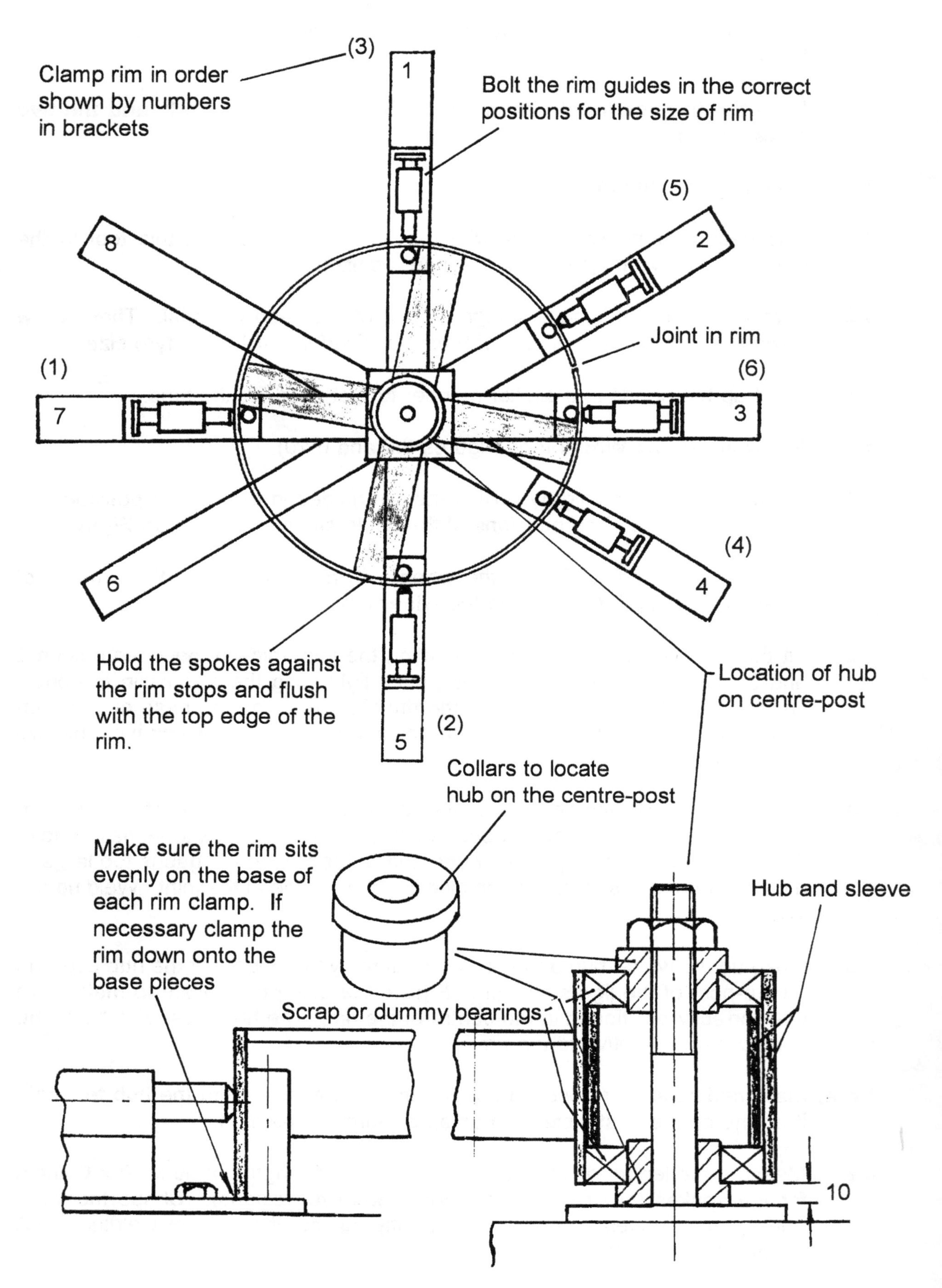


**Figure 6.9: ASSEMBLY OF SPLIT-RIM WHEEL**

**Note:** it is important to cut the spokes accurately to length. If they are too long and are forced into position, the centre-post may be bent. If they are too short, extra welding will be needed to fill the gap and this may distort the rim.

**4.7** Hold the spokes in the correct positions and use strong tack welds to fix them in place.

**4.8** Remove the rim from the jig - note that it may be very tight against the stops and may need to be levered off, or the rim guides to be loosened. Complete the welding around the ends of the spokes. **Note** that any welding on the edge of the rim will need to be ground flush so that the two halves of the rim fit closely together.

**4.9** Fit PART B of the rim in the jig and weld up the joint.

**4.10** Cut **4** pieces of **50 x 50 x 6** angle **40mm** long and drill them for the clamping bolt. Bolt these in position on the spokes and fit the two parts of the rim together. It is a good idea to fit thin spacers - for example a piece of broken hacksaw blade - between the pieces of angle and spoke so that when the clamping bolts are tightened the two parts of the rim are pulled closely together. (Note that the spacers are removed when the wheel is assembled).

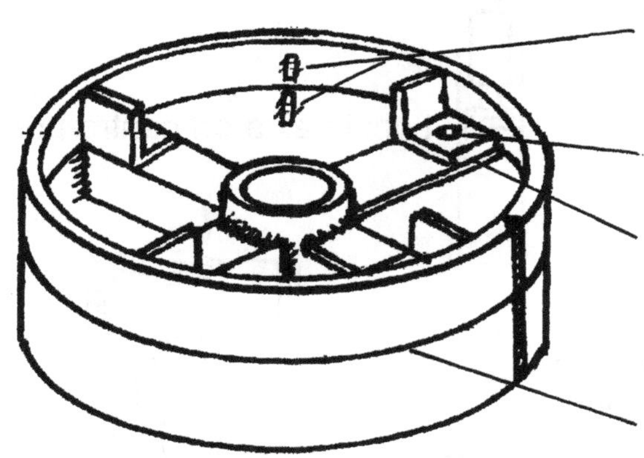


**4.11** Make sure the two parts of the rim fit well together and weld the pieces of angle to PART B of the rim.

**4.12** Since it is likely that the two parts of the wheel will only bolt together in one position it is a good idea to weld pieces of bar inside the rim to mark the correct positioning of the two parts of the wheel.

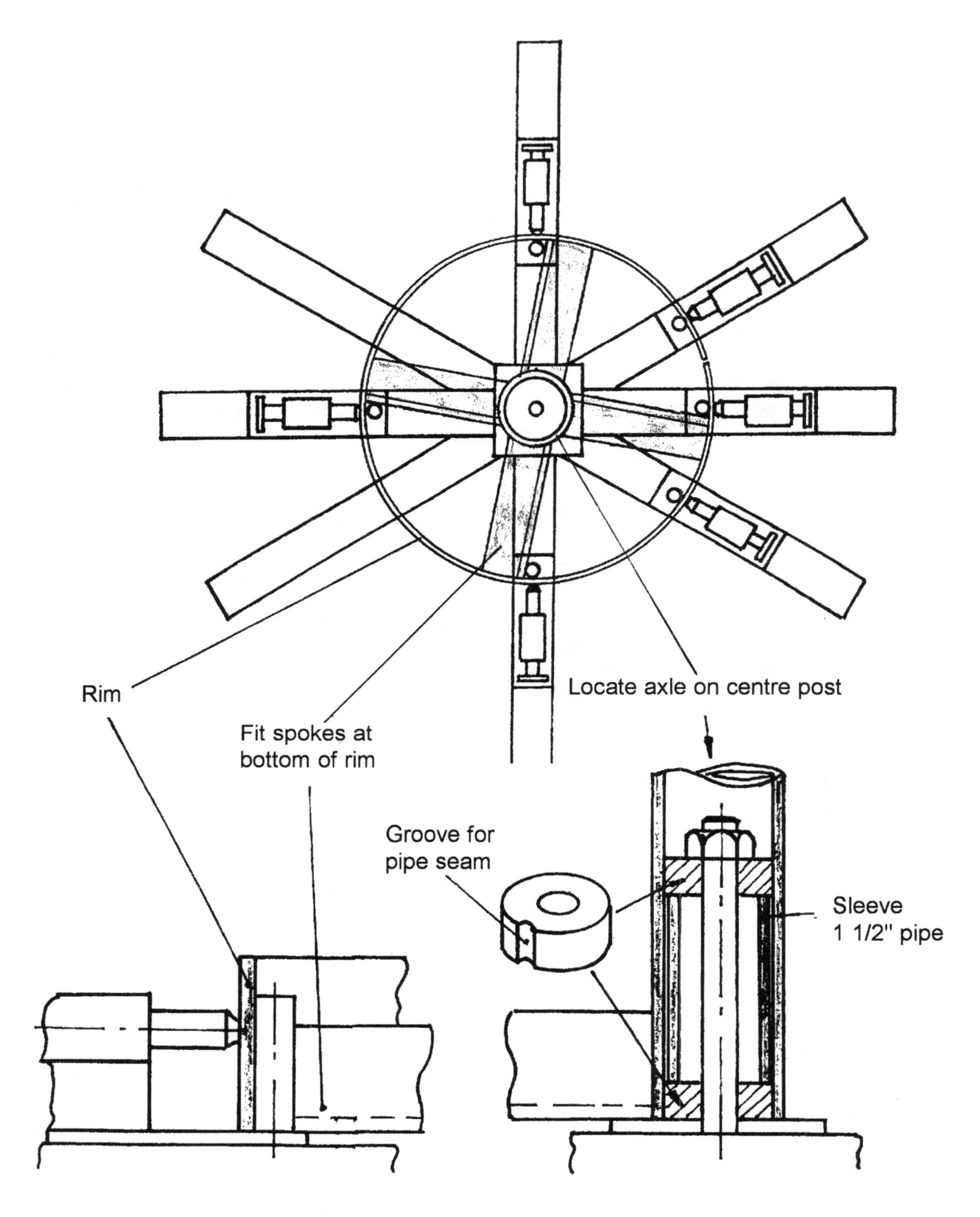


**Figure 6.10: ASSEMBLY OF WHEEL WITH A LIVE AXLE**

**4.13** For each rim part, clamp the bead to the outer edge of the rim and weld it in position with intermittent welds on the outer edge - do not weld inside the bead where the tyre fits. Work around the bead, clamping and welding it in position.

**4.14** Drill a **11mm** diameter hold for the valve in **PART A** of the rim. This should be opposite the joint of the rim.

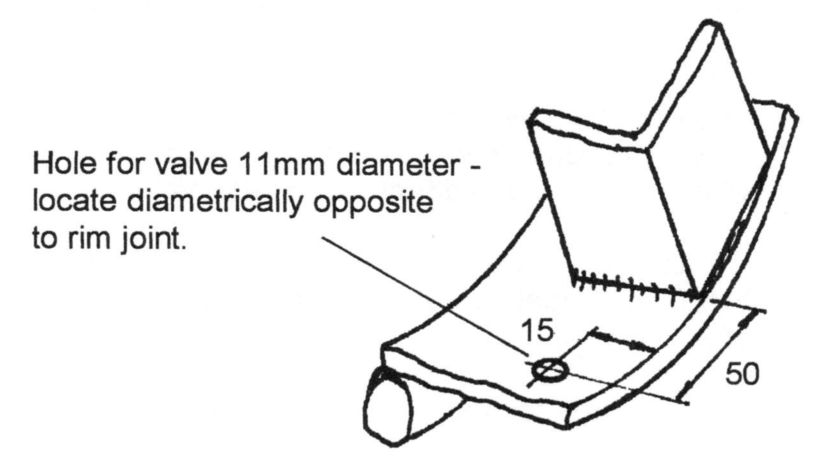


**4.15** Grind/file and smooth off welds and burrs on the outside of the rims that could cause punctures. Pay particular attention to the joint and to any weld splatter.

## 5. Assembly of the Tyre

Fit the tyre and tube on PART A of the rim and slide PART B into the tyre making sure the inner tube is not trapped between the two parts of the rim. Line up the two parts and tighten the joining bolts securely.

The tyre should be inflated to a pressure of about 25 psi - do not over inflate.

**CAUTION: Before loosening the joining bolts ALWAYS make sure the tyre is FULLY DEFLATED.**

### 6.3.2 Detachable-bead wheel

The construction of this wheel follows basically the same steps as for the split-rim wheel. The main steps are illustrated in the following sequence of photographs which will also help to explain the construction of the split-rim wheel.

The differences from the steps set out for the split-rim wheel in 6.3.1 are as follows:

**Rim:** only one rim is needed. This is bent from one length of 100mm wide x 5 or 6mm thick flat bar **or** 2 lengths of 5 or 6 thick bar which give a total width of 100 to 105mm.

If 2 lengths are used, tack-weld them side by side at the ends and bend them together. Weld them together on the **inside** of the rim after bending.

**Spokes:** 6 spokes are needed cut from 50 x 50 x 6 angle.

**Beads:** only 1 bead is welded to the rim. The other is welded to 6 pads for bolting on to the main part of the wheel. The assembly of the bolt-on (detachable) bead is the same as that of PART B of the split-rim wheel.

**Assembly jig:** because the wheel has 6 spokes, the rim will be clamped at 7 positions. The rim-clamps will be located on arms **1,2,3,4,5,6 and 8** of the assembly jig, with the joint between arms **2 and 3**.

1. **Zero the bender**

- fit forming tool;
- rest **straight** length of rim on lower rollers;
- apply slight downward force on lever arm;
- screw up adjusting screw to just touch stop.

2. **Set adjusting screw**

- find number of turns needed for rim size;
- screw adjusting screw down the required number of turns;
- lock the adjusting screw in position.

3. **Bend the rim**

- start with end of rim resting on furthest roller;
- bend rim by forcing lever down onto stop;
- feed rim through in steps shown by chalk marks and bend rim at each step. **Make sure** lever touches stop at each bend.

**4. Complete bending of rim**

- when rim is formed it will have straight pieces at each end;
- remove pivot pin and lever arm - lift rim out of bender.

**5. Form ends of rim**

- hammer the ends of the rim over an anvil or former to make a smooth ring.

**6. Fit rim in assembly jig**

- set jig up for the size of rim and number of spokes;
- clamp rim in jig - locate joint at correct position;
- weld up the rim joint;
- use machined bushes to locate hub on the centre post.

**7. Weld in the spokes**

- measure and cut spokes to length;
- hold them next to rim clamps and flush with top edge of rim;
- weld spokes to rim and hub. **Weld spokes in correct order** of opposite pairs to minimise distortion.

**8. Bend the beads**

- cut round bar for the 2 beads;
- rest a piece of straight bead material on the lower rollers and zero the bender;
- screw the adjusting screw down the required number of turns;
- bend the 2 beads.

**9. Weld on the fixed bead**

- clamp the bead around the outer edge of the rim;
- weld the bead to the rim on the **outside only**;
- cut the overlapping ends of the bead and weld the joint;
- complete welding the bead in position.

10. **Make up the detachable bead**

(or the other part of the rim for a split-rim wheel).

- clamp the bead around the rim, cut the overlapping ends and weld the joints;
- cut pieces of 50 x 6 flat bar and drill clearance holes for M10 bolts;
- hold pieces of flat in position on spokes and weld them to the bead. Weld fully on both sides;
- drill through holes to make bolt holes in spokes;
- clean up the two parts by grinding the welds to make sure the two parts **bolt flush** together;
- drill a hole for the valve.

11. **Fit the tyre**

- fit tyre over rim;
- fit inner tube inside tyre making sure it does not get trapped between the tyre and rim;
- bolt the bead (or other part of rim) **firmly** in position to hold the tyre;
- inflate the tyre.

## 6.4 CONSTRUCTION OF A DETACHABLE-BEAD WHEEL TO TAKE A MOTORCYCLE TYRE

The design of the wheel is shown in Figure 6.11. The wheel is designed to take a 17" or 18" motorcycle tyre, new or used, and has a load capacity of up to 200kg. It is intended for use on heavy duty handcarts or light donkey carts.

Other options for the hub-axle assembly outlined in Section 6.1 may be used instead of the ball bearing assembly shown. If a live axle is used, 1 1/2" medium wall pipe should be used for the axle.

The method of construction of the wheel is exactly the same as described in Section 6.3, and the same steps should be followed:

Some particular notes on the design and construction of the wheel are given below:

**Rim:** the rim diameters and lengths of bar needed are as follows:

| Tyre Size | Outside Diameter of Rim | Length of bar for Rim | Length of bar for Bead |
|---|---|---|---|
| 17" | 426mm | 1,325mm | 1,595mm |
| 18" | 452mm | 1,404mm | 1,675mm |

The rim should be made from 50 or 60mm wide flat bar of 3 to 5mm thickness.

**Spokes:** 6 angle section spokes are used. The preferred size is 30 x 30 of 3,4 or 5 thickness but 25 x 25 x 5 may also be used.

**Axle:** the axle size is 25mm diameter. It may be cut from 25 diameter shafting (bright mild steel bar) or machined from 30 or 32 diameter round bar. Stub axles, 150 long, are welded into each end of an axle-beam of 40 x 40 x 5 angle. Do not weld near the top and bottom surfaces of the axle.

**Bearings:** deep-groove ball bearings type 6205 2RS are used.

**Hub:** this is made from 2" medium-wall pipe. Cut a slot 4 to 5mm wide along the length of the hub, taking out the seam, squeeze up the pipe so that the bearings are a tight fit (use a length of axle to line up the bearings) and weld up the slot. Alternatively the hub may be machined from thick-wall tube or from a fabricated ring.

**Spacer Sleeves:** a sleeve cut from 1" pipe is fitted on the axle between the bearings to prevent the bearings from being loaded when the end nut is tightened. A sleeve cut from 1 1/2" pipe is plug-welded inside the hub to locate the bearings in the hub. This sleeve should be about 1mm shorter than the bearing spacing sleeve (1" pipe).

**Lubrication:** the 6205 2RS bearings are sealed and pre-lubricated. If unsealed bearings are used a grease inlet will be needed in the hub. Drill a 9 or 11 diameter hole mid-way between two spokes, weld on an M8 or M10 nut and use a bolt to plug the inlet. Pack the unsealed bearings with grease during assembly.

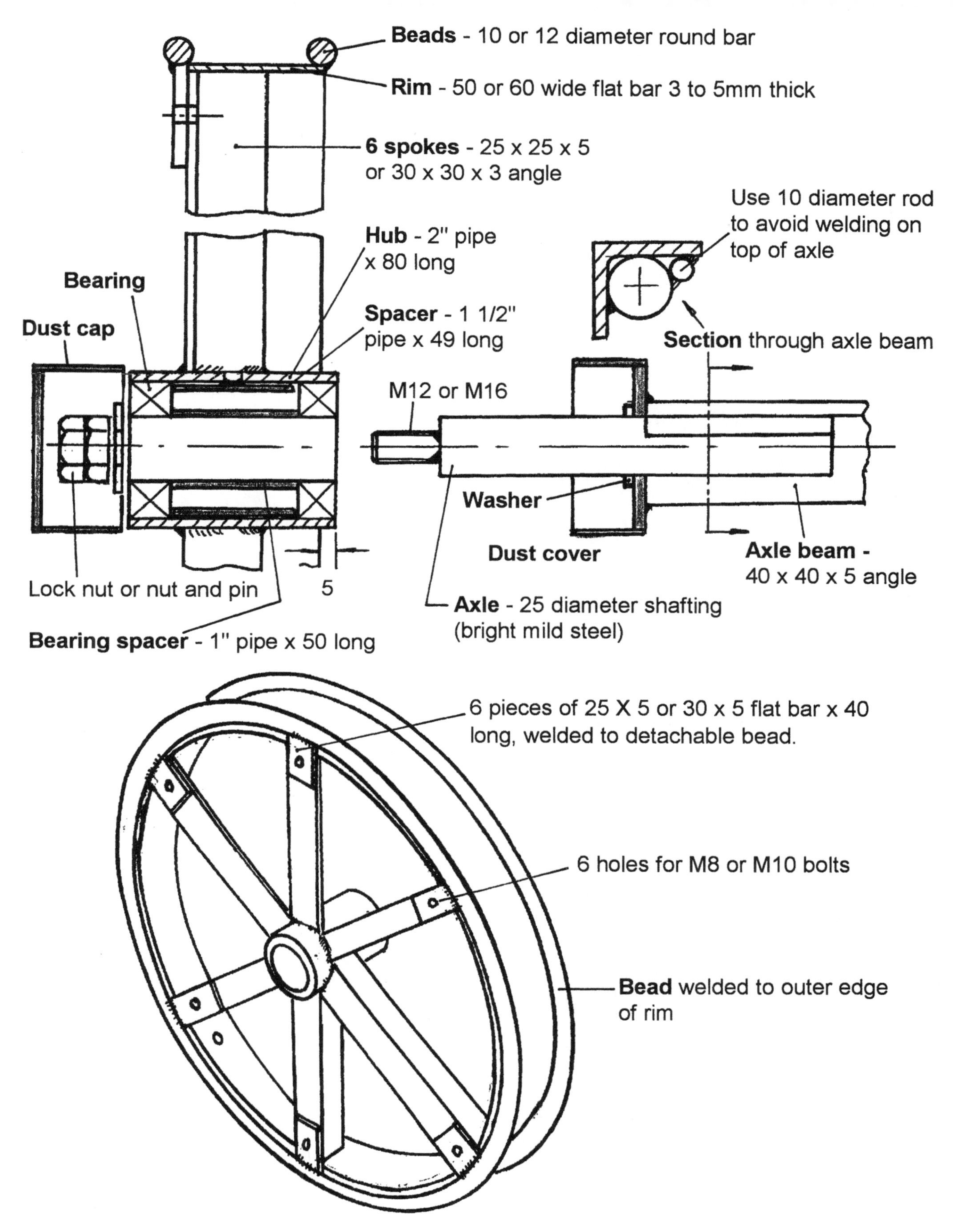


**Figure 6.11: DETACHABLE BEAD WHEEL TO TAKE A MOTORCYCLE TYRE**

# APPENDIX 1

## CONSTRUCTION OF A HAND-OPERATED MACHINE FOR BORING OUT HARDWOOD BEARING BLOCKS

## 1. Details of construction

The design of the components and the assembly of the boring machine are shown in the following drawings. The important features are outlined below:

**Spindle and spindle-guide:** it is very important that the axis of the boring tool lines up accurately with the axes of the spindle and guide and that the spindle is a close fit in the guide, otherwise the boring tool will 'wobble' from side to side when it rotates and will not cut an accurate hole. Therefore, three accurately fitting sizes are needed for the shank of the boring tool, the spindle and spindle-guide. Some possibilities are:

- The best solution is to machine the parts on a lathe. 40 or 50 diameter round bar may be used for the guide, bored out to take a 30 or 32 diameter spindle. The spindle may be drilled to take a 20 or 22 diameter shank

- If a lathe is not available it may be possible to find sizes of round bar and pipe which fit closely together, although this will give less satisfactory performance. 1" medium-wall pipe may be used for the guide and 3/4" medium-wall pipe for the spindle. The 3/4" pipe may be drilled out to take a 22 diameter shank, or it may be possible to find a piece of 5/8" pipe which fits closely enough in the 3/4" pipe to use as a shank

- An alternative method of making a guide is shown in the drawing. This uses 2 lengths of 25 x 25 angle with pieces of flat bar welded to their edges. These can be bolted together with accurately filed spacing pieces between them to give a close running fit of the spindle in the guide.

**Hand-wheel:** this is fixed to the spindle by a locking-screw which clamps on to a flat filed on the shank of the spindle. The hand-wheel may be fitted above or below the lever arm depending which is most convenient for operation. If it is fitted above, it needs to be turned over so that the locking-screw still locates on the flat on the spindle.

**Boring tools:** the cutting blade should be accurately cut and filed from a piece of 6mm thick flat bar. It is then welded centrally into a slot cut in the end of the shank. It is particularly important that the point of the tool is lined up accurately with the axis of the shank. The cutting edges need to be case-hardened and sharpened.

The blade diameters for various axle sizes are as follows:

| Axle size | Cutting diameter `D´ |
|---|---|
| 1" pipe | 34.5 to 35mm |
| 1 1/2" pipe | 49.4 to 50mm |
| 2" pipe | 61.5 to 62mm |

## 2. Operation of the boring machine

A weight is hung from the outer end of the lever arm and the hand wheel turned. The weight should be the largest that can be applied that will still produce a smooth boring operation and a clean cut. A load of about 10kg, producing about 30kg force on the boring tool, seems to be about right for hardwoods.

The machine may also be used for drilling small holes (up to 10mm diameter) in steel, although this is rather slow and tedious. In this case a larger load is needed, about 25 to 30kg at the end of the lever, producing 75 to 90kg on the drill. The drill will need to be fixed centrally in a sleeve of the same diameter as the shank of the boring tool.

## 3. Preparation of bearing blocks

The blocks should be cut and planed to size and holes drilled for the clamping bolts. If the block is to be split into two halves it should be cut accurately through the centre before it is bored. The block should be bolted rigidly to the clamping bar, its centre lined up with the point on the boring tool, and the clamping bar bolted tightly through the slots in the base.

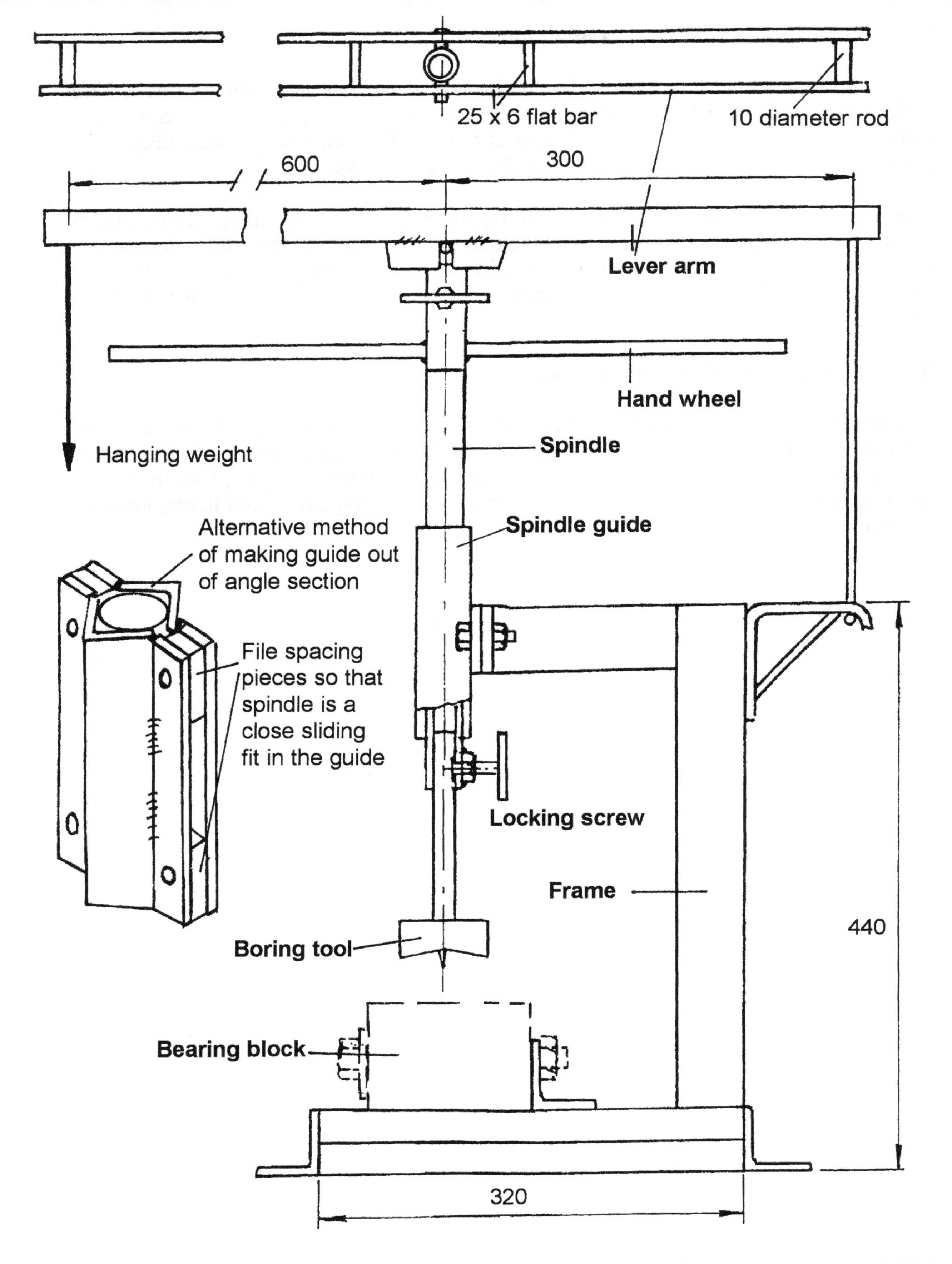


HAND-OPERATED MACHINE FOR BORING OUT HARDWOOD BEARING BLOCKS

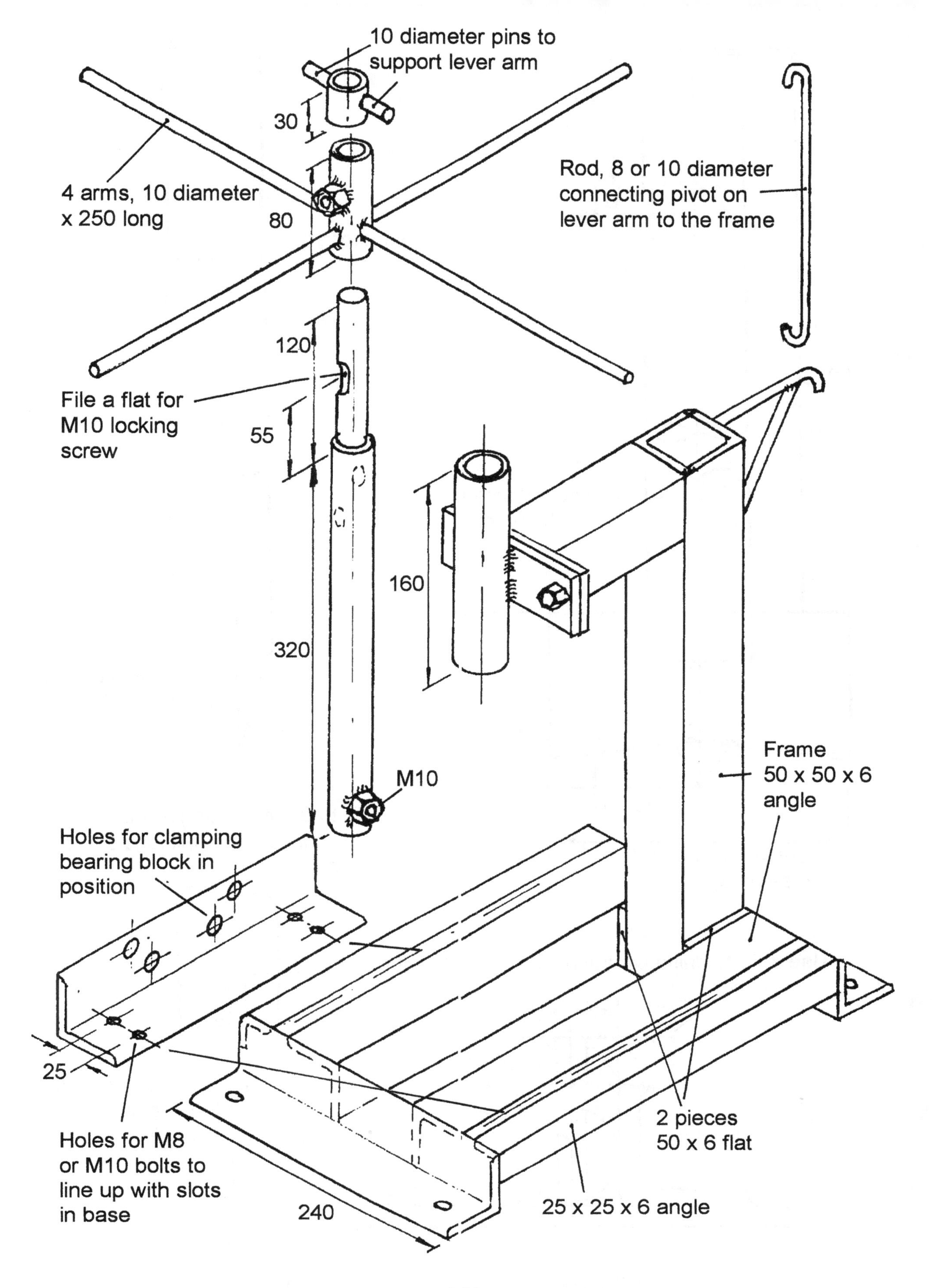
10 diameter pins to
support lever arm
30
4 arms, 10 diameter
x 250 long
80
Rod, 8 or 10 diameter
connecting pivot on
lever arm to the frame
120
File a flat for
M10 locking
screw
55
160
320
Frame
50 x 50 x 6
angle
M10
Holes for clamping
bearing block in
position
25
2 pieces
50 x 6 flat
Holes for M8
or M10 bolts to
line up with slots
in base
240
25 x 25 x 6 angle

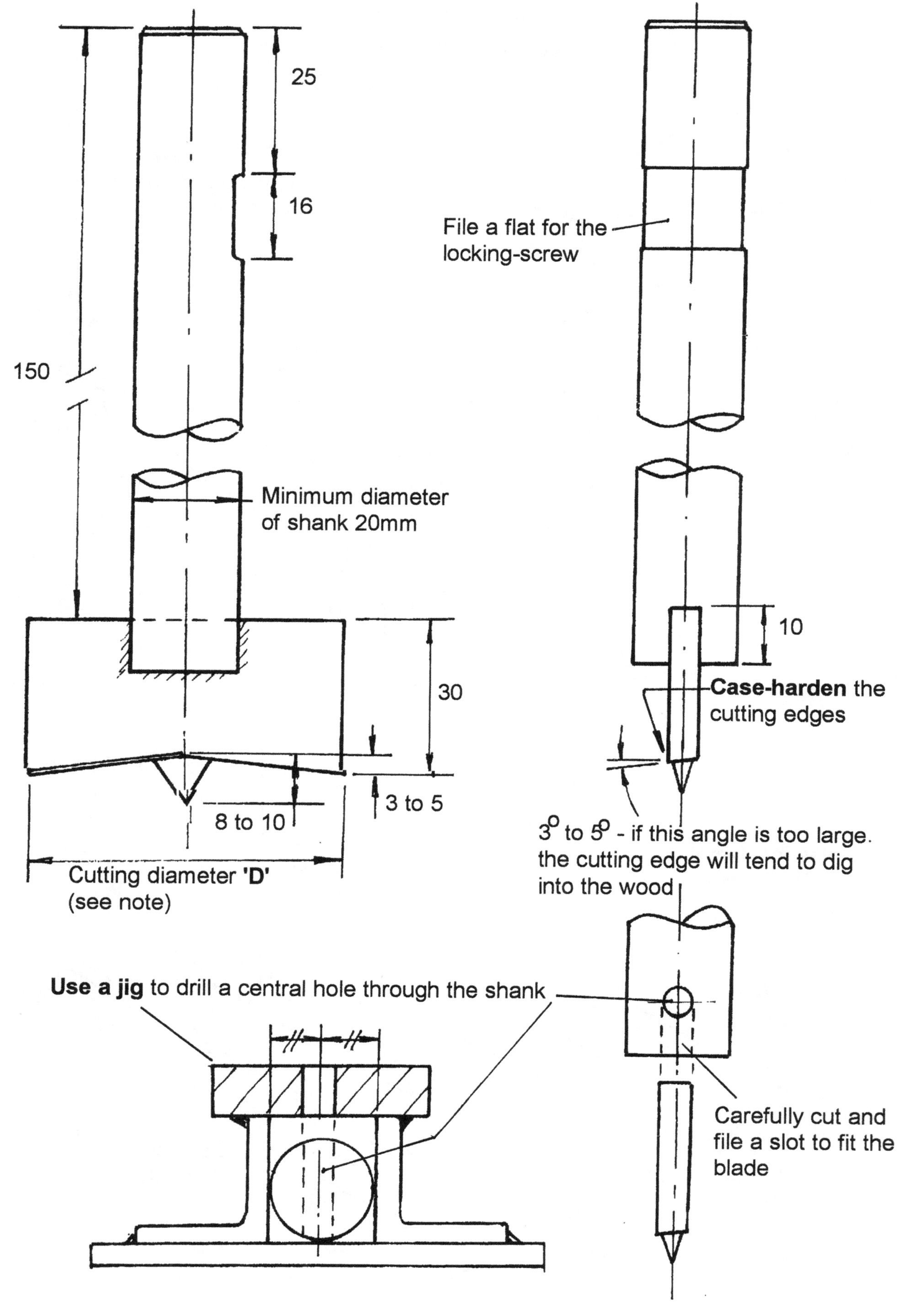


**CONSTRUCTION OF BORING TOOL**

# APPENDIX 2

## TECHNICAL DRAWINGS OF THE RIM-BENDING MACHINE

(Larger, full-page A3-size copies of the drawings in Appendix 2 are available from I.T. Transport Ltd., The Old Power Station, Ardington, Nr. Wantage, Oxon OX12 8QJ, UK.)

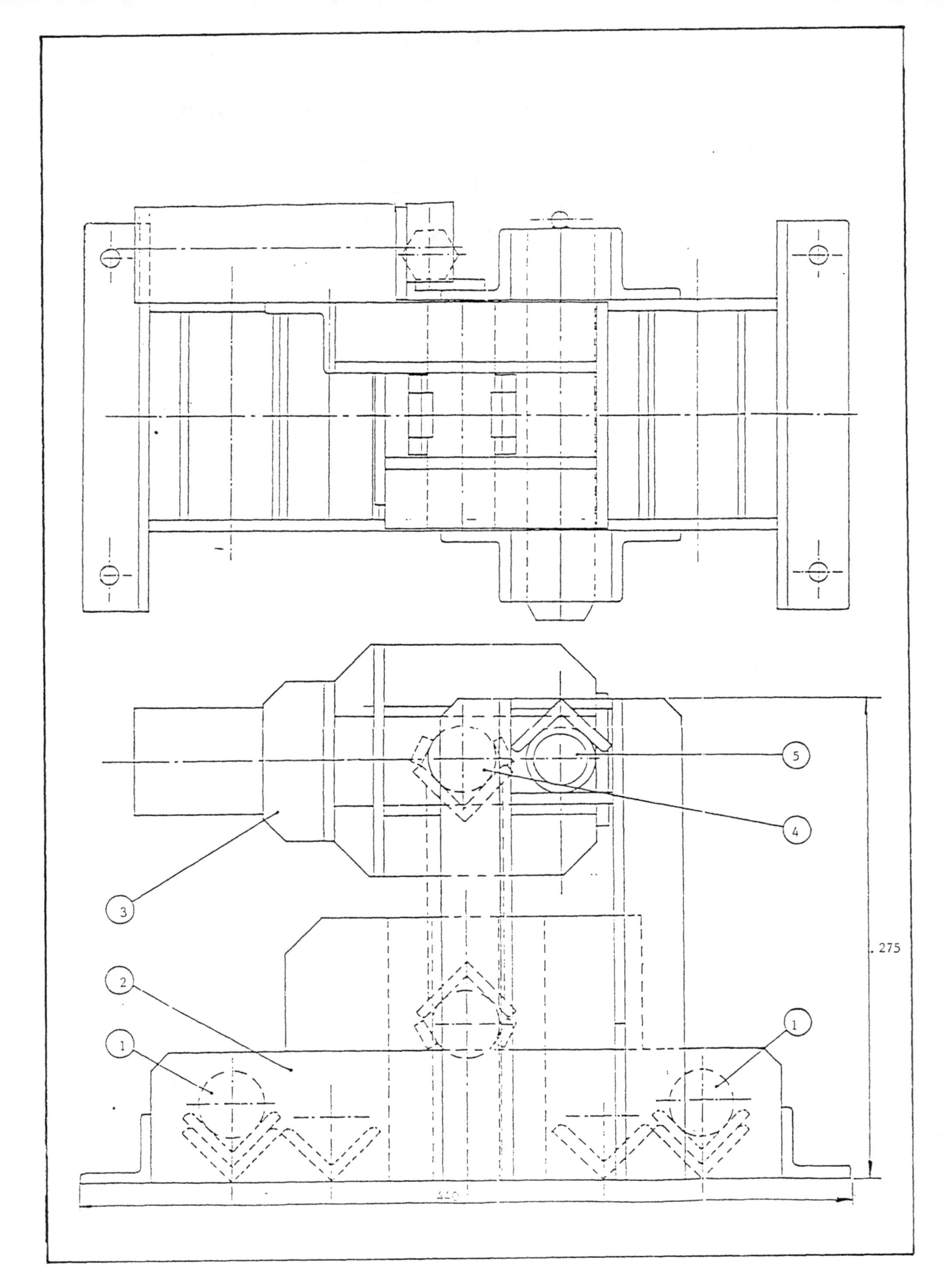
5
4
3
275
2
1
1

I.T. TRANSPORT LTD

TITLE: RIM BENDING DEVICE

PART: GENERAL ASSEMBLY

DATE: 12-7-91 | SCALE: 1:2 | DRAWN BY: R.A.D.

PROJECTION: | DRG. NO: L07/1/100

ALL DIMENSIONS IN MM

| PART Nº | Nº OFF | SPECIFICATION | DRWG |
|---|---|---|---|
| 1 | 2 | BOTTOM ROLLERS | 1/103 |
| 2 | 1 | MAIN FRAME | 1/101 |
| 3 | 1 | LEVER ARM | 1/102 |
| 4 | 1 | TOOL HOLDER PIVOT PIN | 1/103 |
| 5 | 1 | LEVER ARM PIVOT PIN | 1/103 |
| 6 | 3 | FORMING TOOLS | 1/103 |
| 7 | 1 | TOOL HOLDER | 1/103 |
| 8 | 1 | LOCK NUT | |
| 9 | 1 | ADJUSTABLE STOP | 1/101 |
| | | | |

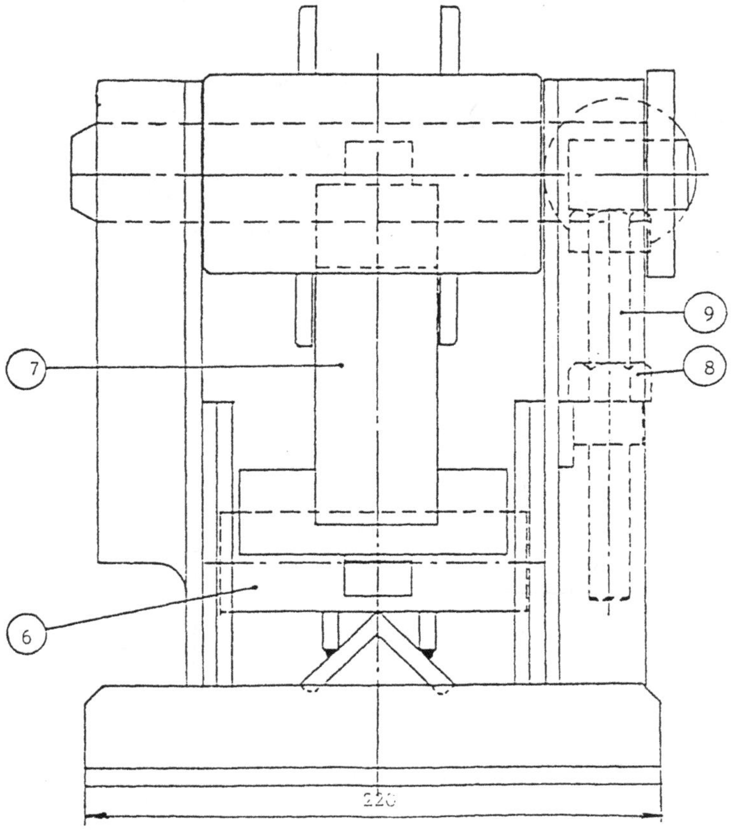

7

6

9

8

220

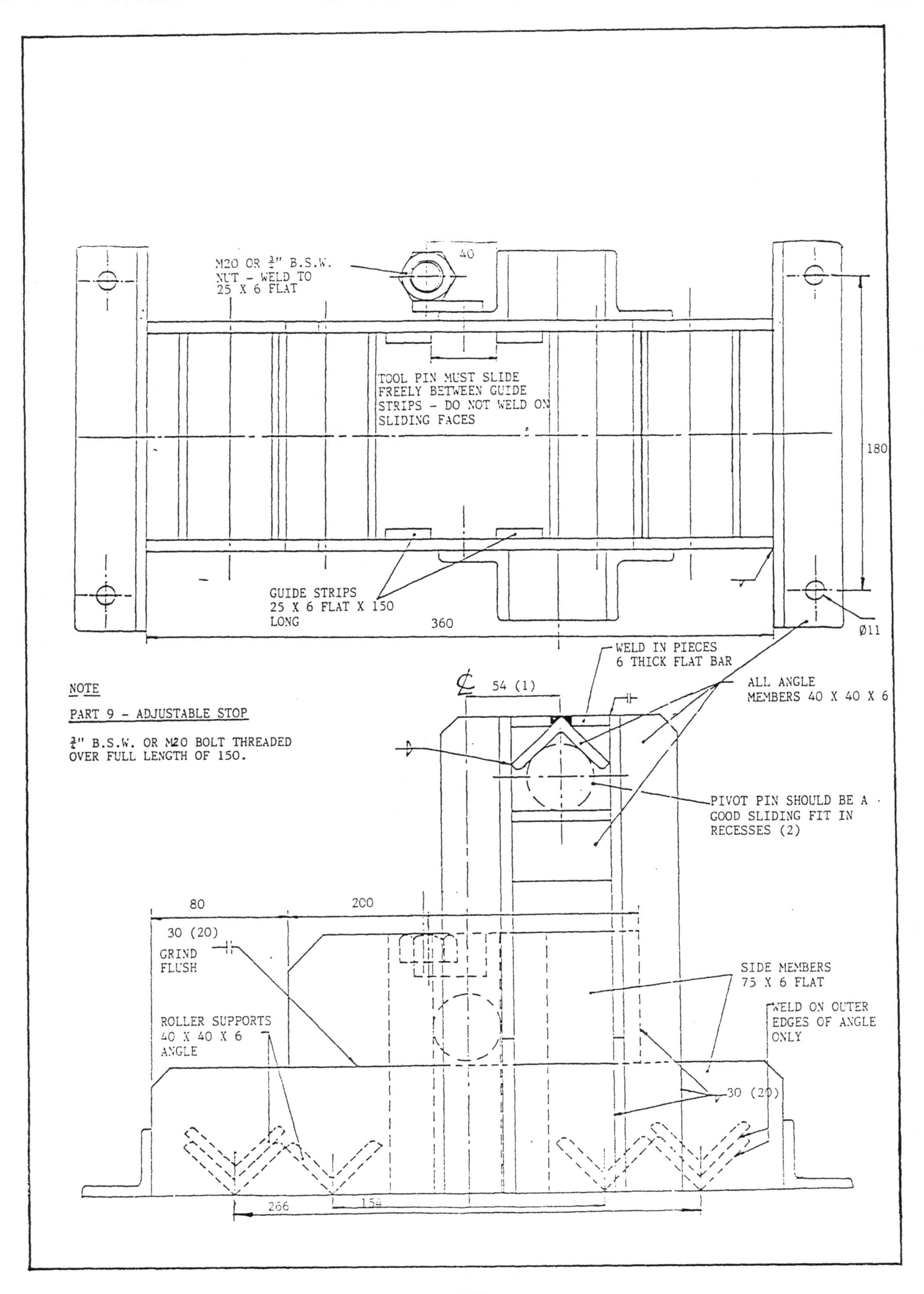
M20 OR ¾" B.S.W.
NUT - WELD TO
25 X 6 FLAT
40
TOOL PIN MUST SLIDE
FREELY BETWEEN GUIDE
STRIPS - DO NOT WELD ON
SLIDING FACES
180
GUIDE STRIPS
25 X 6 FLAT X 150
LONG
360
Ø11
WELD IN PIECES
6 THICK FLAT BAR
NOTE
PART 9 - ADJUSTABLE STOP
¾" B.S.W. OR M20 BOLT THREADED
OVER FULL LENGTH OF 150.
54 (1)
ALL ANGLE
MEMBERS 40 X 40 X 6
PIVOT PIN SHOULD BE A
GOOD SLIDING FIT IN
RECESSES (2)
80
200
30 (20)
GRIND
FLUSH
SIDE MEMBERS
75 X 6 FLAT
WELD ON OUTER
EDGES OF ANGLE
ONLY
ROLLER SUPPORTS
40 X 40 X 6
ANGLE
30 (20)
266
154

| I.T. TRANSPORT LTD | | |
|---|---|---|
| TITLE: RIM BENDING DEVICE | | |
| PART: 2 - MAIN FRAME | | |
| DATE: 12-7-91 | SCALE: 1:2 | DRAWN BY: R.A.D. |
| PROJECTION: | | DRG. NO: L07/1/101 |
| ALL DIMENSIONS IN MM | | |

MATERIAL: MILD STEEL SECTIONS AS INDICATED.

NOTES: WELDS IMPORTANT FOR STRENGTH ARE INDICATED ON DRAWING - OTHER WELDS NOT CRITICAL.

(1) THIS DIMENSION IS IMPORTANT REFERENCE FOR LAYING OUT THE FRAME. IF Ø40 PINS USED, INCREASE CENTRE DISTANCE TO 56MM.

(2) IT IS IMPORTANT THAT AXES OF ALL PINS AND ROLLERS ARE PARALLEL TO ENSURE EVEN BENDING ACTION.

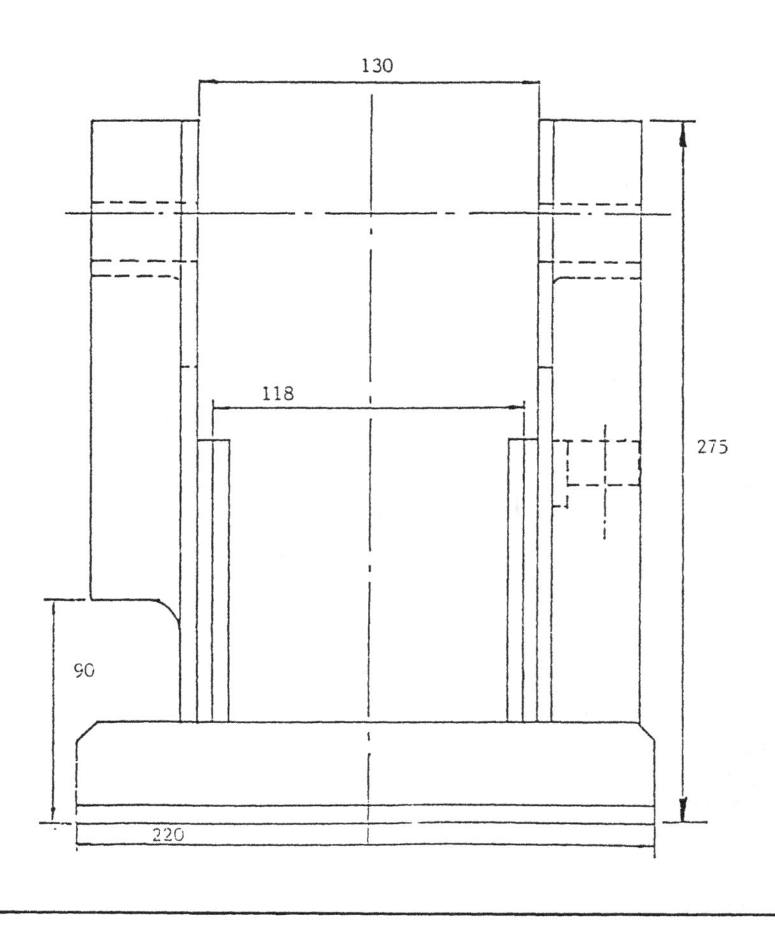

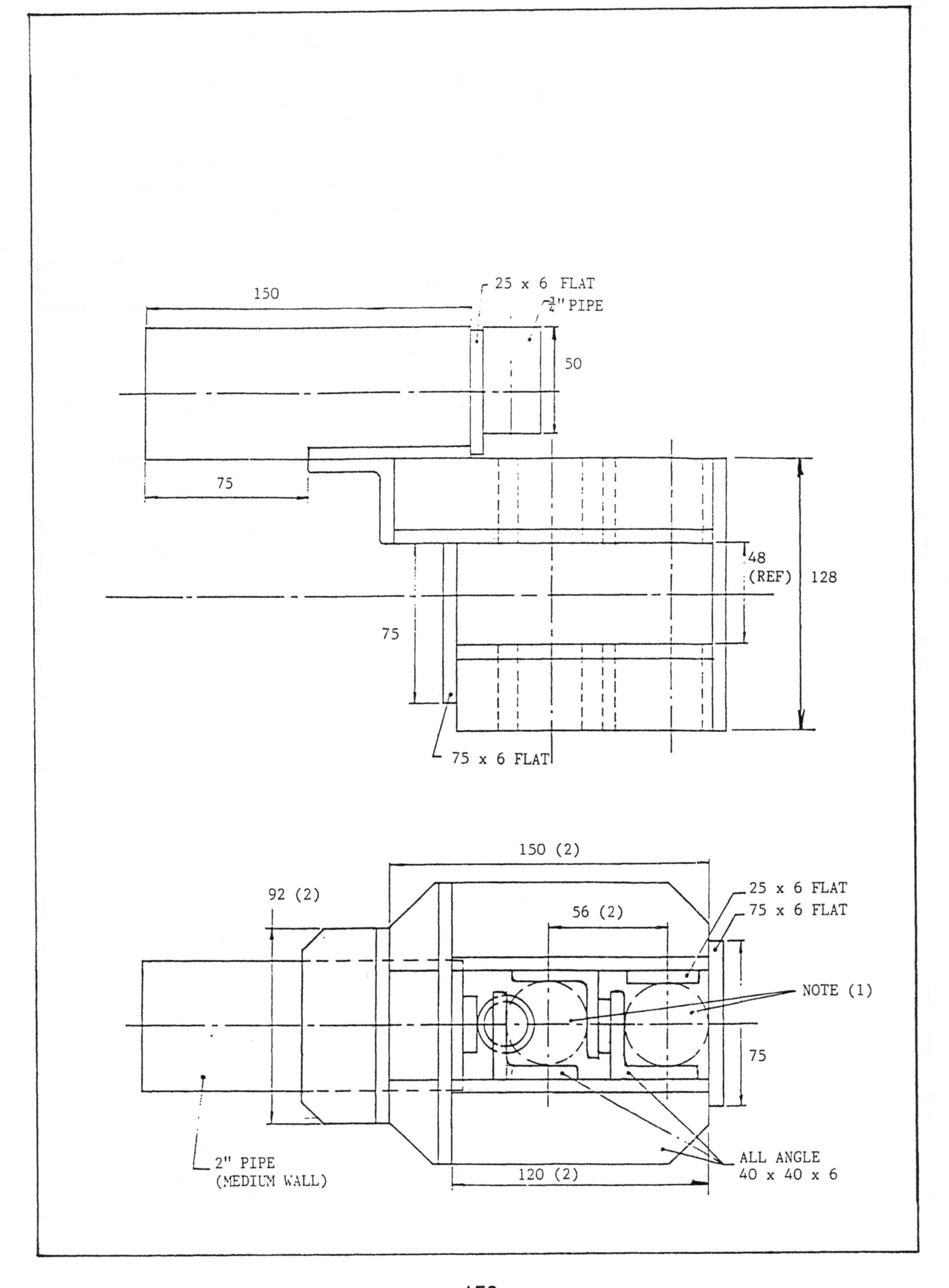

150
25 x 6 FLAT
3/4" PIPE
50
75
48
(REF)
128
75
75 x 6 FLAT
150 (2)
92 (2)
25 x 6 FLAT
75 x 6 FLAT
56 (2)
NOTE (1)
75
2" PIPE
(MEDIUM WALL)
120 (2)
ALL ANGLE
40 x 40 x 6

| I.T. TRANSPORT LTD | | |
|---|---|---|
| TITLE: RIM BENDING DEVICE | | |
| PART: 3 - LEVER ARM | | |
| DATE: 15-7-91 | SCALE: 1:2 | DRAWN BY: R.A.D. |
| PROJECTION: | | DRG. NO: L07 1/102 |
| ALL DIMENSIONS IN MM | | |

MATERIAL: MILD STEEL SECTIONS AS INDICATED.

WELDING : ALL JOINTS SHOULD BE WELDED ALL AROUND AS FAR AS POSSIBLE WITH CONTINUOUS WELDS.

NOTES:
(1) PINS SHOULD BE SQUARE AND PARALLEL AND BE GOOD SLIDING FITS IN RECESSES.

(2) IF Ø40 PINS USED, ADD ON 2MM TO EACH OF THESE DIMENSIONS.

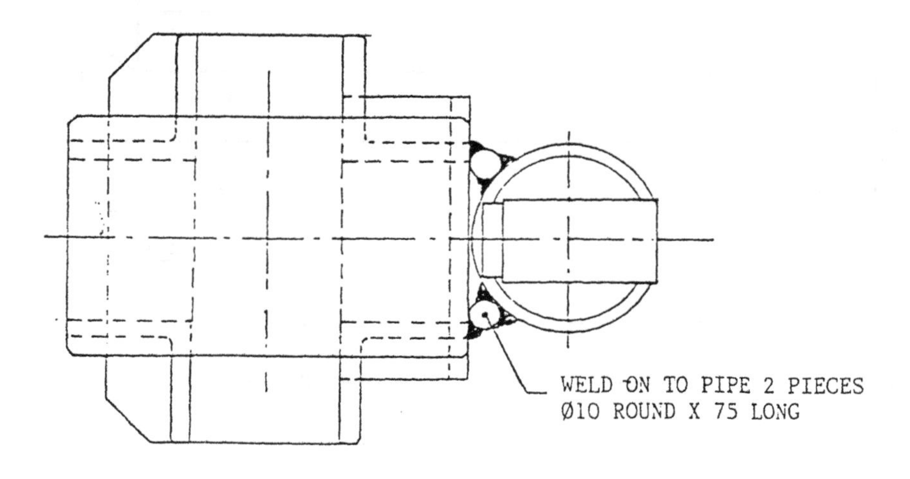

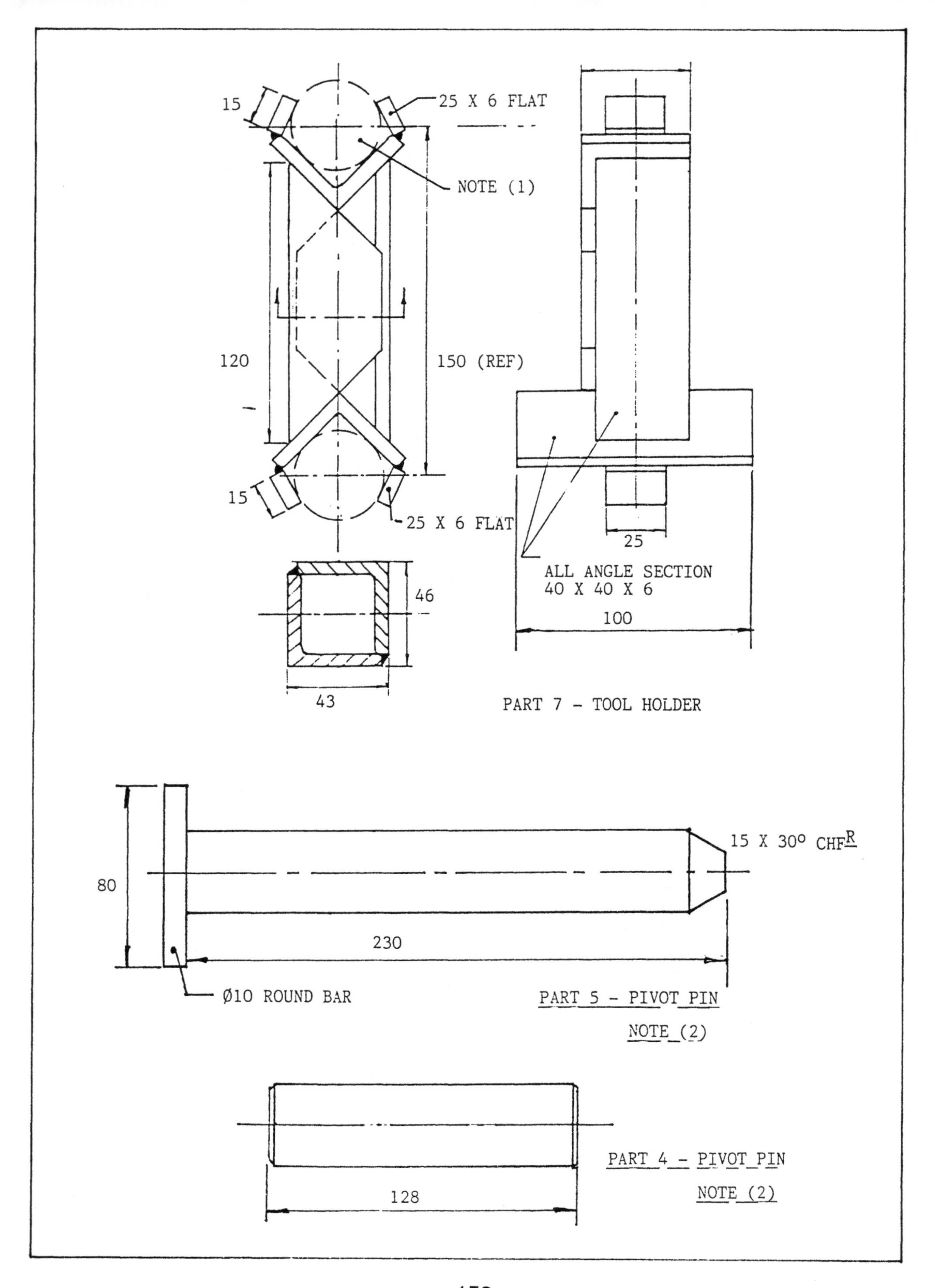

15
25 X 6 FLAT
NOTE (1)
120
150 (REF)
15
25 X 6 FLAT
25
ALL ANGLE SECTION
40 X 40 X 6
46
100
43
PART 7 - TOOL HOLDER
15 X 30° CHFR
80
230
Ø10 ROUND BAR
PART 5 - PIVOT PIN
NOTE (2)
PART 4 - PIVOT PIN
NOTE (2)
128

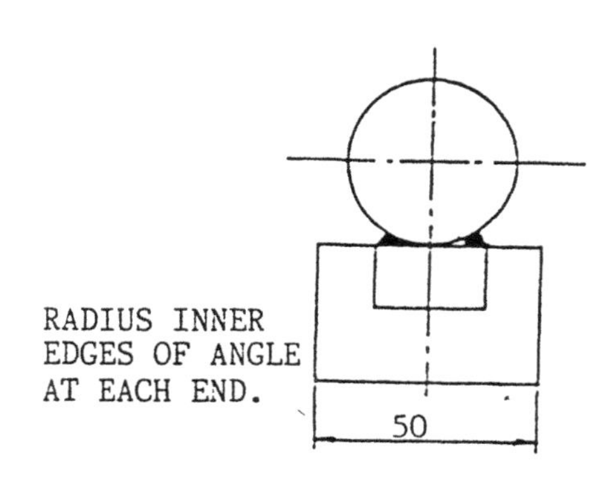


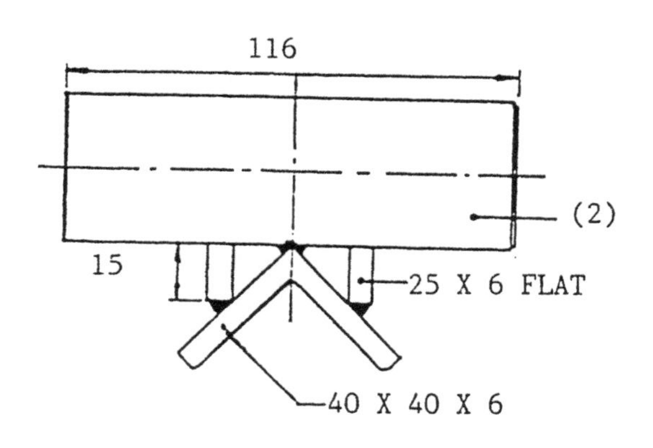


PART 6 - 'V' TOOL

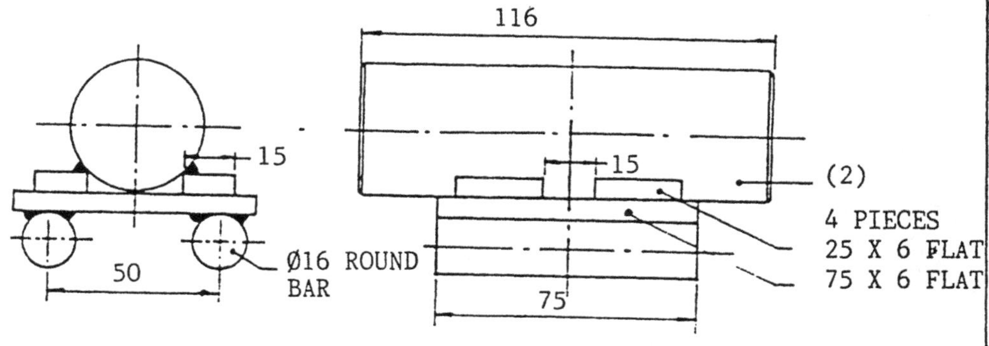


PART 6 - 2 POINT BENDING TOOL

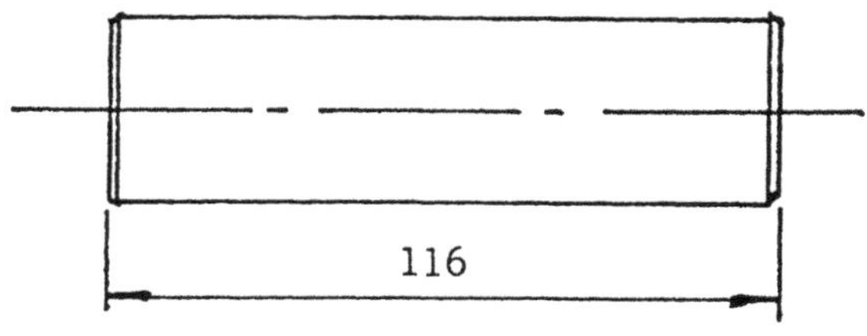


3 REQUIRED: PART 6 - SINGLE POINT BENDING TOOL (NOTE 2)

PART 1 - 2 BOTTOM ROLLERS

NOTES

(1) PINS SHOULD BE PARALLEL AND GOOD SLIDING FITS IN RECESSES.

(2) ALL PINS CUT FROM Ø38 COLD-ROLLED ROUND BAR.

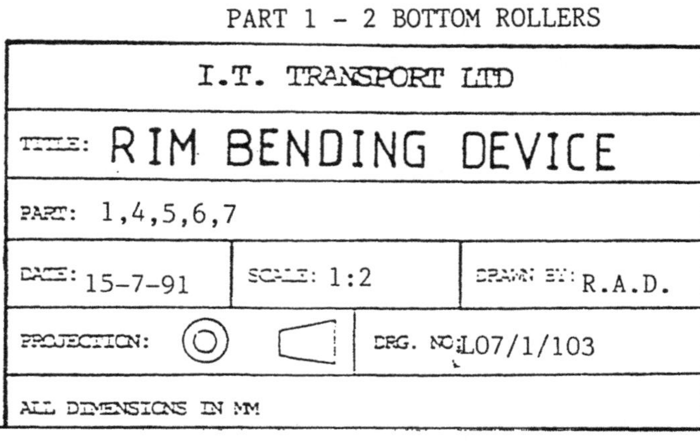

I.T. TRANSPORT LTD

TITLE: RIM BENDING DEVICE

PART: 1,4,5,6,7

DATE: 15-7-91 | SCALE: 1:2 | DRAWN BY: R.A.D.

PROJECTION: | DRG. NO: L07/1/103

ALL DIMENSIONS IN MM

www.ingramcontent.com/pod-product-compliance
Lightning Source LLC
Jackson TN
JSHW060308140125
77033JS00021B/620
* 9 7 8 1 8 5 3 3 9 1 4 1 5 *